UNIFYING CONCEPTS IN GRANULAR MEDIA AND GLASSES

UNIFYING CONCEPTS IN GRANULAR MEDIA AND GLASSES

Edited by

A. Coniglio
A. Fierro
H.J. Herrmann
M. Nicodemi

Università di Napoli "Federico II"
Dipartimento di Scienze Fisiche
Napoli, Italy

ELSEVIER
2004
Amsterdam – Boston – Heidelberg – London – New York – Oxford
Paris – San Diego – San Francisco – Singapore – Sydney – Tokyo

ELSEVIER B.V.
Sara Burgerhartstraat 25
P.O. Box 211, 1000 AE Amsterdam
The Netherlands

ELSEVIER Inc.
525 B Street, Suite 1900
San Diego, CA 92101-4495
USA

ELSEVIER Ltd
The Boulevard, Langford Lane
Kidlington, Oxford OX5 1GB
UK

ELSEVIER Ltd
84 Theobalds Road
London WC1X 8RR
UK

First edition 2004

Library of Congress Cataloging in Publication Data
A catalog record is available from the Library of Congress.

British Library Cataloguing in Publication Data
A catalogue record is available from the British Library.

ISBN: 0 444 51607 7

∞ The paper used in this publication meets the requirements of ANSI/NISO Z39.48-1992 (Permanence of Paper).
Printed in The Netherlands.

Preface

Deep connections are emerging between the physics of non-thermal systems, such as granular media, and other "complex systems" such as glass formers, spin glasses, colloids or gels. The International Workshop "*Unifying concepts in granular media and glasses*", June 25–28, 2003, Villa Orlandi, Capri (Italy), was devoted to the discussion of these topics very important from the point of view of fundamental sciences and for technological applications, ranging from chemistry to fluids mechanics and physics. The conference was mainly sponsored by the SPHYNX Program of the *European Science Foundation* and by Dipartimento di Fisica, *Università di Napoli "Federico II"*.

This book discusses the concepts used for the description of these "complex systems" developed in recent years in the physics community. The special focus of the book is about recent important developments in the formulation of a *Statistical Mechanics* approach *for non-thermal systems*, such as granular media, and the description of out-of-equilibrium dynamics, such as *"jamming" phenomena* ubiquitous in these "complex systems". The book collects contributions from leading researchers in these fields, participating to the workshop, providing both an introduction to these subjects and featuring an up to date presentation of theoretical and experimental developments.

In the last few years important experimental and theoretical discoveries have opened new scenarios concerning the properties of glasses and dense granular materials. These systems exhibit deep similarities in their "jamming" behaviors. Jamming is observed when a constituent particle, surrounded by a crowd of similar particles is strongly constrained in its motion and necessitates large scale rearrangements of many other particles to move. The whole system is thus driven towards a kinetic arrest. One of the most intriguing questions in theoretical physics, today, is whether "jamming" leads to a new state of matter, the "glassy phase", or it is just a situation where flowing is too slow to be observed. The idea of a unified description of "jamming" in different systems, as those mentioned above, is also emerging (see Silbert, O'Hern, Liu and Nagel contribution to this volume), but the precise nature of jamming in non-thermal systems, such as granular media, and the origin of its close connections to glassy phenomena in thermal ones are still open and very important issues. Finally, beyond these general questions, there are many important phenomena specific to groups of these systems, such as size segregation in powders or gelling in colloids, along with their shear thinning and thickening phenomena, to be still understood.

The first part of this book deals with non-thermal systems and, more specifically, granular media. Here there is an even more basic conceptual open problem: the lack of an established theoretical framework where they might be described. Sam Edwards (see Edwards, Brujić, and Makse contribution) proposed a Statistical Mechanics solution to such a problem by introducing the hypothesis that time averages of a system, exploring its mechanically stable states subject to some external drive (e.g., "tapping"), coincide at stationarity with suitable ensemble averages over its "jammed states". In the canonical ensemble, Edwards' distributions are thus characterized by few "thermodynamic parameters", such as "compactivity" or "configurational temperatures". Some recent results investigating and generalizing Edwards' proposal are discussed in this book (see Dean and Lefevre, and Tarjus and Viot contributions). In particular, within Edwards' approach it can be shown that mean field models for granular media undergo a phase transition from a (supercooled) "fluid" phase to a "glassy" phase, when their crystallization transition is avoided (see Nicodemi, Coniglio, de Candia, Fierro, Pica Ciamarra and Tarzia contribution). The nature of such a "glassy" phase results to be the same found in mean field models for glass formers: a discontinuous one step Replica Symmetry Breaking phase preceded by a dynamical freezing point. These results are supported by "tap dynamics" simulations which show a pronounced jamming similar to the one found in experiments on granular media (see Bideau, Philippe, Ribière and Richardet, and Caballero, Lindner, Ovarlez, Reydellet, Lanuza and Clement, and D'Anna and Mayor contributions). As a further application of Edwards' approach to powders, segregation phenomena in these systems are also briefly discussed (see Nicodemi, Coniglio, de Candia, Fierro, Pica Ciamarra and Tarzia) along with their experimental counterpart (see Reis, Mullin and Ehrhardt contribution). Also the role of grains characteristics on the overall system properties is explored (see Mehta and Luck contribution).

The second part of the book deals with thermal glassy systems. The structure of glassy dynamics and aging is discussed and evidence is given about the general possibility to define a dynamical "effective temperature" characterizing off-equilibrium relaxation (see Franz, Lecomte, and Mulet, and Ritort contributions). In particular, the recently discovered glassy properties of dense attractive micellar systems are studied from an experimental (Mallamace, Broccio, Chen, Faraone, and Chen contribution) and a theoretical point of view (see Zaccarelli, Sciortino, Buldyrev, and Tartaglia, and Del Gado, Fierro, deArcangelis, and Coniglio contributions), along with the properties of supercooled water (see Giovambattista, Mazza, Buldyrev, Starr and Stanley contribution). The important idea is also discussed that different phenomena, such as shear thinning, shear thickening and jamming in colloids, can be given a unified description by schematic mode coupling theories (see Cates, Holmes, Fuchs and Henrich contribution).

Summarizing, even though the general validity of Edwards' approach to non-thermal systems has just begun to be assessed, it turns out that a first reference framework is emerging to understand the physics of granular media. At the same time, the nature of their deep connections with "jamming" phenomena in thermal systems, such as glass formers (ranging from colloids and gels to spin glasses) starts being understood. Finally, we wish to warmly thank all the participants to the workshop, and the contributors to the present volume, for their active and stimulating presence.

Antonio Coniglio, Annalisa Fierro, Hans J. Herrmann, Mario Nicodemi
Napoli, June 30th, 2003

Volume Contents

Unifying Concepts in Granular Media and Glasses
A. Coniglio, A. Fierro, H.J. Herrmann and M. Nicodemi (editors)

The Properties of Jamming at Zero Temperature

Leonardo E. Silbert[a b], Corey S. O'Hern[ab c], Andrea J. Liu[a] and Sidney R. Nagel[b]

[a]Department of Chemistry and Biochemistry, UCLA, Los Angeles, CA 90095-1569

[b]James Franck Institute, The University of Chicago, Chicago, IL 60637

[c]Department of Mechanical Engineering, Yale University, New Haven, CT 06520-8284.

We review the idea of the jamming phase diagram that relates jamming in athermal systems such as granular media, foams, and emulsions to glass formation in supercooled liquids. We discuss the properties of a special point on this diagram that occurs at zero temperature and zero applied shear stress. We show using finite-size scaling that this point is well defined in the large system-size limit and that our algorithm for creating these configurations provides a well-defined meaning for the idea of randomness in random close-packing. Near this point, the systems have properties that in some ways are like critical behavior and in other ways are different from what is normally expected at a critical point. The density of normal modes in a system near this point also has unusual features in that it no longer has a regime that is governed by long wavelength plane waves.

Many different systems can flow but become rigid when certain control parameters are varied. For example, a liquid can turn into an amorphous solid by increasing the density or lowering the temperature below the glass-transition temperature; a foam, granular material or colloidal suspension can develop a non-zero shear modulus if the density is increased or if the applied shear stress is lowered below the yield stress. It is tempting to think that the ways in which these different systems develop their rigidity can be related to one another[1].

In an attempt to see how this might be the case, we have proposed a "jamming phase diagram"[2]. In such a diagram, as shown in Figure 1, there are three axes - temperature, T; inverse packing fraction (or density), $1/\phi$; and shear stress, Σ. Systems outside the jamming surface in this diagram can flow, whereas those near the origin – with sufficiently low temperature, high packing fraction and low shear stress so that they lie within the jamming surface – are jammed and have a non-zero long-time shear modulus.

The jamming surface is defined by when the response of the fluid becomes sufficiently slow (i.e., the relaxation time becomes sufficiently long) so that no appreciable flow would be observed by a patient experimentalist. This is the standard definition for the glass-transition temperature [3], since it has never been clear whether or not a true thermodynamic transition occurs between a liquid and a glass. (However, dielectric data on many glass-forming liquids have been interpreted as evidence for such an underlying phase transition[4].) Despite this ambiguity, it is nevertheless clear that a dramatic slowing occurs

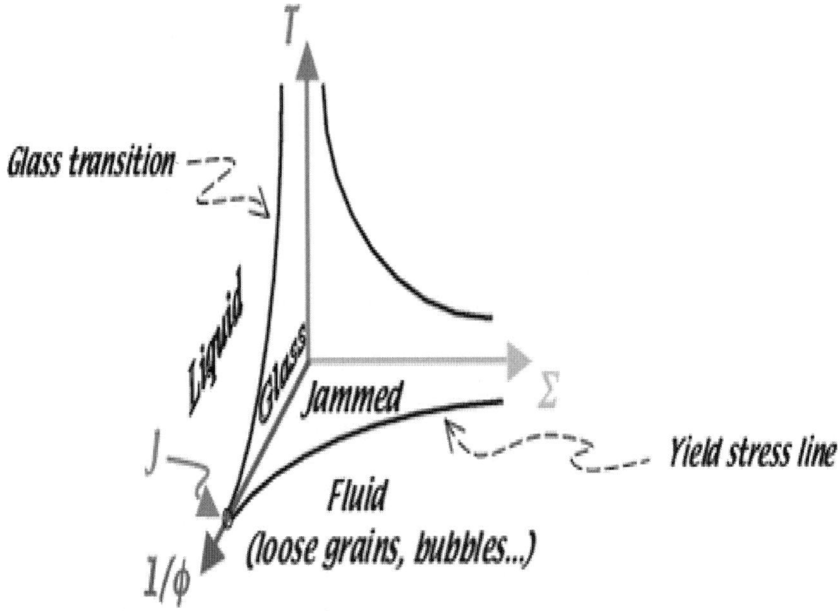

Figure 1. A schematic jamming phase diagram. The point marked "J" exists for finite ranged repulsive potentials and marks the point at $T = 0$ and $\Sigma = 0$ where jamming takes place along the $1/\phi$ axis.

as the liquid is cooled or compressed towards the glass state or as the shear stress is decreased toward the jamming threshold. As the control parameters are varied, the system can pass through the jamming surface. The ordinary glass transition would occur in the vertical plane (T, $1/\phi$) coming out of the paper. The horizontal plane ($1/\phi$, Σ) coming out of the paper corresponds to the ordinary jamming transition where systems can flow if the shear stress exceeds the yield stress. The glass-transition and the yield-stress lines are marked on the diagram. Weitz et al. [5] have shown that such a generic phase diagram is useful for correlating features of the fluid-to-solid transition even for attractive colloidal particles.

There is a special point on the diagram, marked "J", where the yield-stress and the glass-transition lines meet at $T = 0$ and $\Sigma = 0$ along the $1/\phi$ axis. Such a point occurs for finite-range, repulsive potentials. Near the point of contact, such potentials have the form:

$$V(r) = \begin{matrix} V_0(\sigma - r)^{\alpha} & \quad r < \sigma \\ 0 & \quad r \geq \sigma \end{matrix} \tag{1}$$

where σ is the particle diameter and the exponent $\alpha = 2.0$ and $\alpha = 2.5$ corresponds to harmonic springs and Hertz potentials respectively.

We have shown that point J has a number of special properties [6,7]. Because the jamming surface is defined by where the relaxation time reached some arbitrary large but finite value, the surface is not sharply defined because its position will depend on the chosen relaxation-time criterion. We have shown that point J, on the other hand, is well defined in the large system-size limit and that at this point, the infinite-time bulk and shear moduli simultaneously become non-zero on increasing the packing fraction.

For finite-size systems we have found that there is a distribution of packing fractions, ϕ_c, at which jamming occurs at $T = 0$ and $\Sigma = 0$. This distribution becomes narrower and approaches a delta-function as the number of particles, N, approaches infinity. In Figure 2, we show the finite-size scaling results for the onset of jamming thresholds as N is varied. Periodic boundary conditions are used and we have studied bidisperse systems in 2 dimensions and both monodisperse and bidisperse systems in 3 dimensions. The bidisperse systems are $50 - 50$ mixtures of particles with diameters σ and 1.4σ. We start the simulation at a fixed packing fraction, ϕ, at infinite temperature so that the initial particle positions can be chosen completely at random. We then quench the system with conjugate gradient techniques to $T = 0$ to find the inherent structures[8].

If there are any particle overlaps at low packing fraction, the potentials will create forces so that the particles will shift until the overlaps are completely eliminated. At some packing fraction ϕ_c the complete elimination of overlaps is no longer possible and the system begins to develop a non-zero potential energy and a non-zero pressure. The fraction of configurations that are jammed by this criterion at a packing fraction, ϕ, are shown in Figure 2a for a 2-dimensional system of bi-disperse particles. Each curve is for a different system size N. For $N = 16$ the onset of jamming is quite gradual as the packing fraction is increased. As N grows the onset becomes very sharp. The distribution of thresholds, ϕ_c, are shown in Figure 2b for the same set of configurations as shown in Figure 2a.

In Figure 2c, we show the full width at half maximum of these distributions. Data for 3-dimensional systems as well as for two-dimensions is included. The width narrows as N increases and varies approximately as a power law going to zero as N goes to infinity. This indicates that in the infinite-system-size limit, all systems (for a given dimension, dispersity and potential) jam at the same packing fraction, ϕ^*. The approach to ϕ^* as N increases is shown in Figure 2d.

These results show that Point J is well defined in the large N limit. Moreover, we find that for 3-dimensional, monodisperse systems, the value of the packing fraction ϕ^* at which jamming occurs in the large system-size limit is very close to the value normally associated with random close-packing: $\phi^* = 0.64$. We have found that ϕ^* appears to be independent of the exponent α. Indeed, the distributions of jamming thresholds are, within the numerical uncertainty, identical for potentials with $\alpha = 2.5, 2.0$, and 1.5. This result strongly suggests that the same value for ϕ^* would be associated with the $\alpha = 0$ (hard-sphere) limit for the potential. The algorithm (based on inherent structures quenched from infinite temperature at fixed ϕ - i.e., in a fixed potential-energy landscape) that we have used to produce the configurations at ϕ_c, provides a well-defined definition of what "random" means for random close-packing or, equivalently, provides a definition for "maximally random jammed" [9]. That is, it shows that essentially *all* initial configurations jam at the same packing fraction in the large system-size limit.

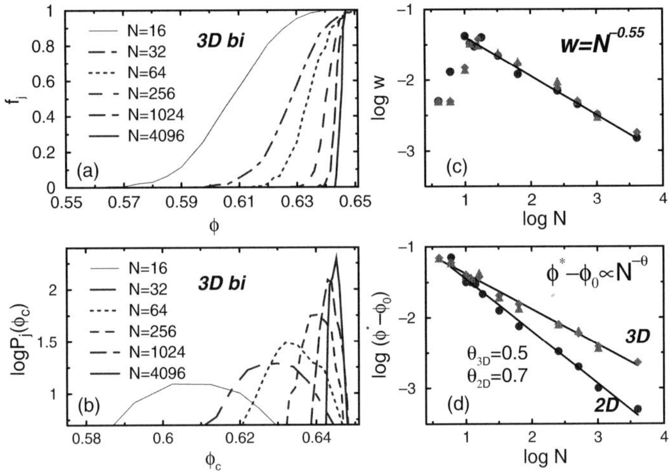

Figure 2. Finite-size scaling results for the onset of jamming. (a) The fraction of configurations that are jammed as the system is quenched from infinite temperature to $T = 0$ for a 3-dimensional bidisperse system. Each curve corresponds to a different value of N, the number of particles in the system. (b) The distribution of jamming thresholds, $P_j(\phi_c)$ as a function of ϕ_c for the same distributions shown in (a). (c) The full width at half maximum for $P_j(\phi_c)$ for 2- (blue) and 3- (red) dimensional systems with both harmonic and Hertzian potentials. (d) The position of the peak in $P_j(\phi_c)$ versus N. At large N, the distribution approaches a delta function centered at ϕ^* where ϕ^* is, for monodisperse systems in 3-dimensions, the value associated with random close-packing.

As we mentioned above, the shear modulus, G, becomes non-zero at the same value of the packing fraction, ϕ_c, as does the pressure, p (or bulk modulus). This is shown in Figure 3 for both 2- and 3- dimensional systems with harmonic ($\alpha = 2$) and Hertz ($\alpha = 2.5$) potentials. Here both p and G scale as a power of $(\phi - \phi_c)$ and both quantities go to zero at the same value of ϕ_c. Thus these system jam at the same value of ϕ both in the sense that the particles begin to have unavoidable overlap with one another and in the sense that the system develops a shear modulus.

We have also shown that point J corresponds to an isostatic state where the average number of overlaps per particle, Z, in the connected cluster is equal to the number of constraints required for mechanical stability. For frictionless spheres this corresponds to $Z = Z_c = 2D$ where D is the dimension of the system. As shown in Figure 3c, the average number of overlapping neighbors, Z, increases above the isostatic value, Z_c, as a power of $(\phi - \phi_c)$. This exponent is independent of both the dimension and the potential.

We have found that point J has many properties that resemble a critical point: (i) As shown in Figure 2, there is finite-size scaling of the distribution of jamming thresholds. (ii) As shown in Figure 3a and 3b, the pressure, p, and shear modulus, G, increase as power-laws versus $(\phi - \phi_c)$. (iii) There is also a power-law divergence in the first peak of the pair-distribution function, $g(r)$ as ϕ approaches ϕ_c. (iv) There is a lack of self-

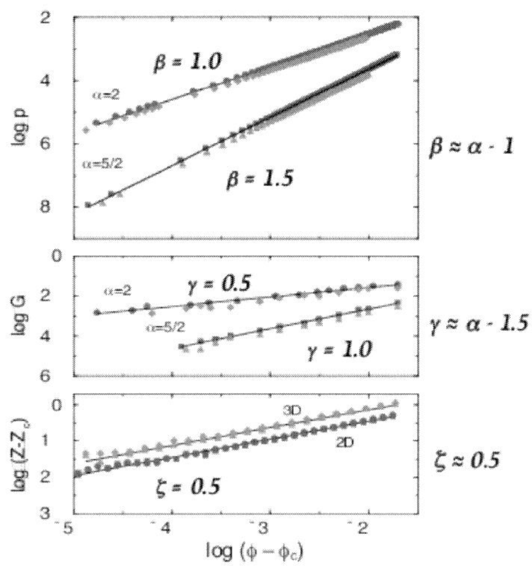

Figure 3. Scaling of (a) the pressure, p, (b) the shear modulus, G, and (c) the excess number of overlapping neighbors, $(Z - Z_c)$, as a function of $(\phi - \phi_c)$. The pressure and shear modulus both become non-zero at the same value of ϕ_c. All three quantities scale as a power of $(\phi - \phi_c)$. In all the curves the blue (red) points corresponds to data in 2-(3-) dimensions. The exponents in the scaling relations for the pressure, β, and the shear modulus, γ, are independent of dimension but depend on the potential of interaction: $\beta \approx \alpha - 1$; $\gamma \approx \alpha - 1.5$. The scaling exponent for $(Z - Z_c)$ is independent of dimension and potential.

averaging that appears in the distribution of forces as the correlation length exceeds the size of the system near ϕ_c. (v) As shown in Figure 3c, the average number of overlapping neighbors, Z, increases above the isostatic value, Z_c, as a power of $(\phi - \phi_c)$.

On the other hand, we also found that point J has properties that are very unusual for critical phenomena. The exponents for the pressure and the shear modulus vary with the potential (i.e., the value of α) but do not depend on the dimension D of the system. There is a discontinuous jump in the number of overlapping neighbors from 0 below ϕ_c to Z_c above ϕ_c. There are no fluctuations in the energy, pressure, or shear modulus at packing fractions below ϕ_c. Finally, the fluctuations appear quite differently in constant-pressure (which implies a constant value of $(\phi - \phi_c)$) versus constant-volume ensembles.

Finally we point out that the density of phonon states at point J is also highly unusual. Normally one expects that at sufficiently low frequencies, normal modes will be long-wavelength plane waves. This would produce a density of modes varying as $D(\omega) \propto \omega^2$ (in 3 dimensions). In dramatic contrast, we find that $D(\omega)$ approaches a constant value at zero frequency near point J. This is shown in Figure 4 where $D(\omega)$ is plotted for different values of $(\phi - \phi_c)$.

At higher packing fractions, the density of states resembles that found by earlier researchers on similar systems [10]: $D(\omega)$ rises from zero near $\omega = 0$, approaches a broad

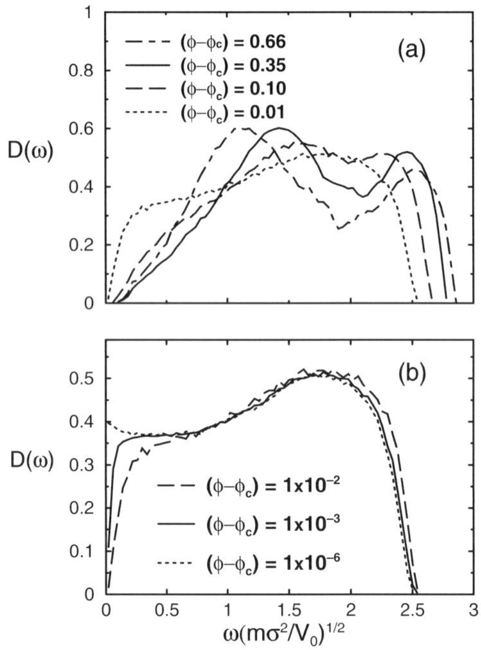

Figure 4. The density of states for a 3-dimensional monodisperse system with harmonic ($\alpha = 2$) interactions. Each curve corresponds to a different value of $(\phi - \phi_c)$. For $(\phi - \phi_c) = 10^{-6}$, $D(\omega)$ approaches a constant value at low ω.

peak and again falls to zero at high ω. On decreasing the packing fraction, the low frequency behavior becomes distorted: the region over which $D(\omega) \propto \omega^2$ becomes narrower until it disappears. At the lowest value of $(\phi - \phi_c)$, we find that the density of states approaches a constant value at low frequency. (This result is consistent with an earlier result on a Lennard-Jones glass in which bonds were randomly cut[11]. As more bonds were cut, so that Z decreased, the density of states at low frequency began to increase as well.) This indicates that the density of states near Point J is as far as possible from what one would expect for a crystal. It suggests that Point J is at the epitome of disorder.

Because the density of states for a system near $(\phi - \phi_c) = 0$ is so different from what one expects in a crystal, it is interesting to examine the low-frequency normal modes themselves. To do this we plot in Figure 5 the eigenvector of the lowest frequency mode for a 2-dimensional, $N = 1024$ system with $(\phi - \phi_c) = 10^{-6}$. There is very little evidence of plane-wave character to this mode. Instead it seems to be composed of patches with swirls of different sizes. This pattern is similar to that seen in more compressed systems at low frequency[12].

In conclusion, we have reviewed the properties of point J in the jamming phase diagram of Figure 1. We note that in many ways, this point behaves like a critical point while in other ways it has features not ordinarily associated with criticality. The density of states

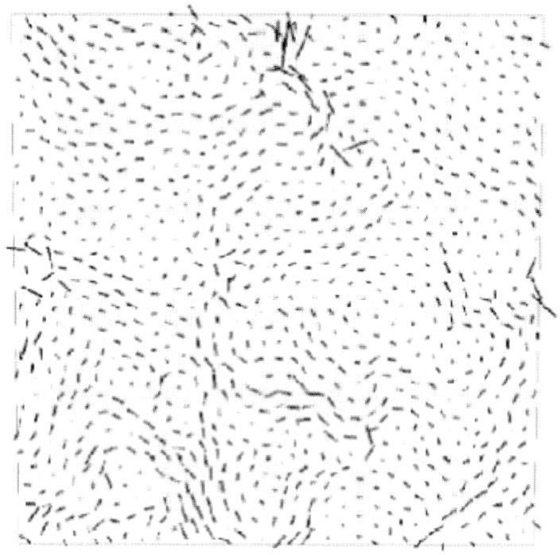

Figure 5. The eigenvector of the lowest frequency mode for a 2-dimensional, $N = 1024$ system with $(\phi - \phi_c) = 10^{-6}$. The length of each line represents the magnitude of the eigenvector at the particle at that position. The line is tangent to the direction of the eigenvector at that point.

and the divergence in the first peak in the pair distribution function, $g(r)$, appear to be relevant to the properties found in real laboratory glasses and it is tempting to think that a complete understanding of the physics at Point J may be a fruitful beginning to understanding the entire jamming phase diagram.

1. Acknowledgments

This work was supported by DOE grant: DE-FG02-03ER46088 (at Univ. of Chicago) and DOE grant: DE-FG02-03ER46087 (at UCLA).

REFERENCES

1. "Jamming and Rheology: Constrained Dynamics on Microscopic and Macroscopic Scales," Edited by A. J. Liu and S. R. Nagel. (Taylor and Francis, London, 2001). ISBN # 0-7484-0879-7.
2. A. J. Liu and S. R. Nagel, "Jamming is not just cool any more," Nature **396**, 21-22 (1998).
3. D. Turnbull, "Under What Conditions can a Glass be Formed?" Contemporary Physics **10**, 473-488 (1969).
4. N. Menon and S. R. Nagel, "Evidence for a Divergent Susceptibility at the Glass Transition," Phys. Rev. Lett. **74**, 1230-1233 (1995).
5. V. Trappe, V. Prasad, L. Cipelletti, P. N. Segre, D. A. Weitz, "Jamming phase diagram for attractive particles," Nature **411**, 772 (2001); V. Prasad, V. Trappe, A. D.

Dinsmore, P. N. Segre, L. Cipelletti, and D. A. Weitz, "Universal features of the fluid to solid transition for attractive colloidal particles", Faraday Discuss. **123**, 1 (2003).
6. C. S. O'Hern, S. A. Langer, A. J. Liu, and S. R. Nagel, "Random Packings of Frictionless Particles," Phys. Rev. Lett. **88**, 075507–1-4 (2002).
7. C. S. O'Hern, L. E. Silbert, A. J. Liu and S. R. Nagel, "Jamming at zero temperature and zero applied stress: the epitome of disorder," Phys. Rev. E '**68**, 011306 1-19 (2003).
8. F. H. Stillinger and T. A. Weber, "Packing structures and transitions in liquids and solids ," Science **225**, 983 (1984); F. H. Stillinger and T. A. Weber, "Dynamics of structural transitions in liquids," Phys. Rev. A **28**, 2408 (1983).
9. S. Torquato, T. M. Truskett, and P. G. Debenedetti, "Is random close packing of spheres well defined?" Phys. Rev. Lett. **84**, 2064 (2000); S. Torquato and F. H. Stillinger, "Multiplicity of generation, selection, and classifcation procedures for jammed hard-particle packings," J. Phys. Chem. B **105**, 11849 (2001); A. R. Kansal, S. Torquato, and F. H. Stillinger, "Diversity of order and densities in jammed hard-particle packings," Phys. Rev. E **66**, 041109 (2002).
10. For example, see: A. Rahman, M. J. Mandell, and J. P. McTague, "Molecular dynamics study of an amorphous Lennard-Jones system at low temperature," J. Chem. Phys. **64**, 1564-1568 (1976).
11. S. R. Nagel, G. S. Grest, S. Feng, and L. M. Schwartz, "Excess Low-Temperature Specific Heat in a Model Glass," Phys. Rev. B **34**, 8667-8670 (1986).
12. J. P. Wittmer, A. Tanguy, J.-L. Barrat, and L. Lewis, "Vibrations of amorphous, nanometric structures: When does continuum theory apply?" Europhysics Lett. **57**, 423-429 (2002); A. Tanguy, J. P. Wittmer, F. Leonforte, and J.-L. Barrat, "Continuum limit of amorphous elastic bodies: A finite-size study of low-frequency harmonic vibrations," Phys. Rev, B **66**, 174205 (2002).

Unifying Concepts in Granular Media and Glasses
A. Coniglio, A. Fierro, H.J. Herrmann and M. Nicodemi (editors)

A basis for the statistical mechanics of granular systems

Sam. F. Edwards[a], Jasna Brujić[a], Hernán A. Makse[b]

[a]Polymers and Colloids Group, Cavendish Laboratory, University of Cambridge, Madingley Road, Cambridge CB3 0HE, UK

[b]Levich Institute and Physics Department, City College of New York,New York, NY 10031, US

This paper aims to justify the use of statistical mechanics tools in situations where the system is out of equilibrium and jammed. Specifically, we derive a Boltzmann equation for a jammed granular system and show that the Boltzmann's analysis can be used to produce a "Second Law", $\dot{S} \geq 0$ for jammed systems. We highlight the fundamental questions in this area of physics and point to the key quantities in characterising a packing of particles, accessible through a novel experimentation method which we also present here.

1. Introduction

In a thermal system, the Brownian motion of the constituent particles implies that the system dynamically explores the available energy landscape, such that the notion of a statistical ensemble applies. For densely packed systems of interest in this study, in which enduring contacts between particles are important, the potential energy barrier prohibits an equivalent random motion. At first sight it seems that the thermal statistical mechanics do not apply to these systems as there is no mechanism for averaging over the configurational states. Hence, these systems are inherently out of equilibrium. On the other hand, if the granular material is gently tapped such that the grains can slowly explore the available configurations, the situation becomes analogous to the equilibrium case scenario. It has been shown that the volume of the system is dependent on the applied tapping regime, and that this dependence is reversible, implying ergodicity [1]. This result gives support to the proposed statistical ensemble valid for dense, static and slowly moving granular materials which was first introduced by Edwards and Oakeshott in 1989 [2,3]. Through this approach, notions of macroscopic quantities such as entropy and compactivity were also introduced to granular matter.

Here we present a theoretical framework to fully describe the exact specificities of granular packings, and a shaking scenario which leads to the derivation of the Boltzmann equation for a jammed granular system. This kind of an analysis paves the path to the study of macroscopic quantities, such as the compactivity, characterising each jammed configuration from the microstructural information of the packing. It is according to this theory that the static configurations obtained from experiments are later characterised. An extended version of this paper is presented in [4].

2. Classical Statistical Mechanics

We first present the classical statistical mechanics theorems to an extent which facilitates an understanding of the important concepts for the development of an analogous granular theory, as well as the assumptions necessary for the belief in such a parallel approach. In the conventional statistical mechanics of thermal systems, the different possible configurations, or microstates, of the system are given by points in the phase space of all positions and momenta $\{p, q\}$ of the constituent particles. The equilibrium probability density ρ_{eqm} must be a stationary state of Liouville's equation which implies that ρ_{eqm} must be expressed only in terms of the total energy of the system, E. The simplest form for a system with Hamiltonian $\mathcal{H}(p, q)$ is the microcanonical distribution:

$$\rho_{\text{eqm}}(E) = \frac{1}{\Sigma_{\text{eqm}}(E)}, \tag{1}$$

for the microstates within the ensemble, $\mathcal{H}(p, q) = E$, and zero otherwise. Here,

$$\Sigma_{\text{eqm}}(E) = \int \delta(E - \mathcal{H}(p, q)) \;\; dp \; dq, \tag{2}$$

is the area of energy surface $\mathcal{H}(p, q) = E$.

Equation (1) states that all microstates are equally probable. Assuming that this is the true distribution of the system implies accepting the ergodic hypothesis, i.e. the trajectory of the closed system will pass arbitrarily close to any point in phase space.

It was the remarkable step of Boltzmann to associate this statistical concept of the number of microstates with the thermodynamic notion of entropy through his famous formula

$$S_{\text{eqm}}(E) = k_B \log \Omega_{\text{eqm}}(E). \tag{3}$$

Whereas the study of thermal systems has had the advantage of available statistical mechanics tools for the exploration of the phase space, an entirely new statistical method, unrelated to the temperature, had to be constructed for grains.

3. Statistical Mechanics for Jammed Matter

We now consider a jammed granular system composed of *rigid* grains. Such a system is analogously described by a network of contacts between the constituent particles in a fixed volume V, since there is no relevant energy E in the system. In the case of granular materials, the analogue of phase space, the space of microstates of the system, is the space of possible jammed configurations as a function of the degrees of freedom of the system $\{\zeta\}$.

It is argued that it is the volume of this system, rather than the energy, which is the key macroscopic quantity governing the behaviour of granular matter [2,3]. If we have N grains of specified shape which are assumed to be infinitely rigid, the system's statistics would be defined by a volume function $\mathcal{W}(\zeta)$, a function which gives the volume of the system in terms of the specification of the grains.

In this analogy one replaces the Hamiltonian $\mathcal{H}(p, q)$ of the system by the volume function, $\mathcal{W}(\zeta)$. The average of $\mathcal{W}(\zeta)$ over all the jammed configurations determines the volume V of the system in the same way as the average of the Hamiltonian determines the average energy E of the system.

3.1. Definition of the volume function, $\mathcal{W}$

One of the key questions in this analogy is to establish the 'correct' $\mathcal{W}$ function, the statistics of which is capable of fully describing the system as a whole. The idea is to partition the volume of the system into different subsystems, α with volume $\mathcal{W}^\alpha$, such that the total volume of a particular configuration is

$$\mathcal{W}(\zeta) = \sum_\alpha \mathcal{W}^\alpha \tag{4}$$

It could be that considering the volume of the first coordination shell of particles around each grain is sufficient; thus, we may identify the partition α with each grain. However, particles further away may also play a role in the collective system response due to enduring contacts, in which case $\mathcal{W}$ should encompass further coordination shells. In reality, of course, the collective nature of the system induces contributions from grains which are indeed further away from the grain in question, but the consideration of only its nearest neighbours is a good starting point for solving the system, and is the way in which we proceed to describe the $\mathcal{W}$ function. The significance of the appropriate definition of $\mathcal{W}$ is best understood by the consideration of a response to an external perturbation to the system in terms of analogies with the Boltzmann equation which we will describe in Section 5.

Ball and Blumenfeld [5] have shown by a triangulation method that the area of the two-dimensional problem can be given in terms of the contact points using vectors constructed from them. Here we consider a cruder version for the volume per grain, yet with a strong physical meaning. For a pair of grains in contact (assumed to be point contacts for rough, rigid grains) the grains are labelled α, β, and the vector from the centre of α to that of β is denoted as $\mathbf{R}^{\alpha\beta}$ and specifies the complete geometrical information of the packing. The first step is to construct a configurational tensor $\mathcal{C}^\alpha$ associated with each grain α based on the structural information,

$$\mathcal{C}^\alpha_{ij} = \sum_\beta R^{\alpha\beta}_i R^{\alpha\beta}_j . \tag{5}$$

Then an approximation for the area in 2D or volume in 3D encompassing the first coordination shell of the grain in question is given as

$$\mathcal{W}^\alpha = 2\sqrt{\mathrm{Det}\mathcal{C}^\alpha_{ij}}. \tag{6}$$

The volume function is depicted in the Fig. 1, with grain coordination number 3 in two dimensions, where Eq. (6) should give the area of the triangle (thick gray lines) constructed by the centres P of grains which are in contact with the α grain. The above equation is exact if the area is considered as the determinant of the vector cross product matrix of the two sides of the triangle. However, this definition is clearly only an approximation of the space available to each grain since there is an overlap of $\mathcal{W}^\alpha$ for grains belonging to the same coordination shell. Thus, it overestimates the total volume of the system: $\sum \mathcal{W}^\alpha > V$. However, it is the simplest approximation for the system based on a single coordination shell of a grain.

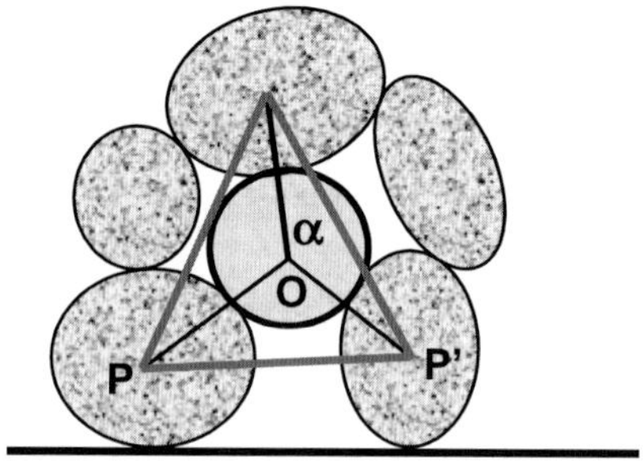

Figure 1. Volume function $\mathcal{W}$ as discussed in the text.

3.2. Entropy and compactivity

Now that we have explicitly defined $\mathcal{W}$ it is possible to define the entropy of the granular packing. The number of microstates for a given volume V is measured by the area of the surface $\mathcal{W}(\zeta) = V$ in the phase space of jammed configurations and it is given by:

$$\Sigma_{\text{jammed}}(V) = \int \delta\big(V - \mathcal{W}(\zeta)\big) \ \ \Theta(\zeta) \ \ d\zeta, \tag{7}$$

where now $d\zeta$ refers to an integral over all possible jammed configurations and $\delta(V - \mathcal{W}(\zeta))$ formally imposes the constraint to the states in the sub-space $\mathcal{W}(\zeta) = V$. $\Theta(\zeta)$ is a constraint that restricts the summation to only reversible jammed configurations [4]. The radical step is the assumption of equally probable microstates which leads to an analogous thermodynamic entropy associated with this statistical quantity:

$$S(V) = \lambda \log \Sigma_{\text{jammed}}(V) = \lambda \log \int \delta(V - \mathcal{W}(\zeta)) \ \ \Theta(\zeta) \ \ d\zeta, \tag{8}$$

which governs the macroscopic behaviour of the system [2,3]. Here λ plays the role of the Boltzmann constant. The corresponding analogue of temperature, named the "compactivity", is defined as

$$X^{-1} = \frac{\partial S}{\partial V}. \tag{9}$$

This is a bold statement, which perhaps requires further explanation in terms of the actual role of compactivity in describing granular systems. We can think of the compactivity as a measure of how much more compact the system could be, i.e. a large compactivity implies a loose configuration (e.g. random loose packing, RLP) while a reduced compactivity implies a more compact structure (e.g. random close packing, RCP). In terms of the reversible branch of the compaction curve obtained in the Chicago experiments [1] large amplitudes generate packings of high compactivities, while in the limit

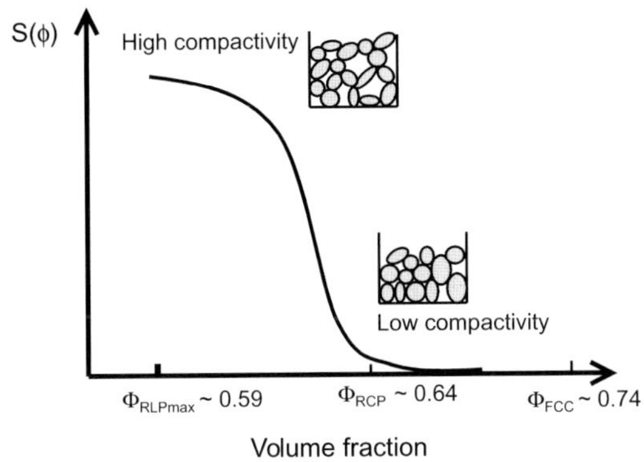

Figure 2. Interpretation of the compactivity and entropy in terms of different packings.

of the amplitude going to zero a low compactivity is achieved. In terms of the entropy, many more configurations are available at high compactivity, thus the dependence of the entropy on the volume fraction can be qualitatively described as in Fig. 2. In the figure, for monodisperse packings the RCP is identified at $\phi \approx 0.64$, the maximum RLP fraction is identified at $\phi \approx 0.59$, while the crystalline packing, FCC, is at $\phi = 0.74$ but cannot be reached by tapping.

3.3. Remarks

To summarise, the granular thermodynamics is based on two postulates:

1) While in the Gibbs construction one assumes that the physical quantities are obtained as an average over all possible configurations at a given energy, the granular ensemble consists of only the jammed configurations at the appropriate volume.

2) As in the microcanonical equilibrium ensemble, the strong ergodic hypothesis is that all jammed configurations of a given volume can be taken to have equal statistical probabilities.

The ergodic hypothesis for granular matter was treated with skepticism, mainly because a real powder bears knowledge of its formation and the experiments are therefore history dependent. Thus, any problem in soil mechanics or even a controlled pouring of a sand pile does not satisfy the condition of all jammed states being accessible to one another as ergodicity has not been achieved, and the thermodynamic picture is therefore not valid. However the Chicago experiments of tapping columns [1] showed the existence of reversible situations. For instance, let the volume of the column be $V(n, \Gamma)$ where n is the number of taps and Γ is the strength of the tap. If one first obtain a volume $V(n_1, \Gamma_1)$, and then repeat the experiment at a different tap intensity and obtain $V(n_2, \Gamma_2)$, when we return to tapping at (n_1, Γ_1) one obtains a volume V' which is $V'(n_1, \Gamma_1) = V(n_1, \Gamma_1)$. There have been several further experiments confirming these results for different system geometries, particle elasticities and compaction techniques, e. g. the system can be mechanically tapped or oscillated, vibrated using a loudspeaker, or even allowed to relax under large pressures over long periods of time, all to the same effect [6–8]. Moreover, in simulations of slowly sheared granular systems the ergodic hypothesis was shown to work [9].

It is often noted in the literature that although the simple concept of summing over all jammed states which occupy a volume V works, there is no first principle derivation of the probability distribution of the granular ensemble as it is provided by Liouville's theorem for equilibrium statistical mechanics of liquids and gases. In granular thermodynamics there is no justification for the use of the $\mathcal{W}$ function to describe the system as Liouville's theorem justifies the use of the energy in the microcanonical ensemble. In Section 5 we will provide an intuitive proof for the use of $\mathcal{W}$ in granular thermodynamics by the analogous proof of the Boltzmann equation.

The comment was nevertheless made that there is no proof that the entropy Eq. (8) is a rigorous basis for granular statistical mechanics. Here we develop a Boltzmann equation for jammed systems and show that this analysis can be used to produce a second law of thermodynamics, $\delta S \geq 0$ for granular matter, and the equality only comes with Eq. (8) being achieved. Although everyone believes that the second law of thermodynamics is universally true in thermal systems, the only accessible proof comes in the Boltzmann equation, as the ergodic theory is a difficult branch of mathematics which will not be covered in the present discussion. By investigating the assumptions and key points which led to the derivation of the Boltzmann equation in thermal systems, it is possible to draw analogies for an equivalent derivation in jammed systems.

It should be noted that there is an extensive literature on granular gases [10,11], which are observed when particles are fluidised by vigorous shaking, thus inducing continuous particle collisions. There is a powerful literature on this topic, but it is not applicable to the problem of jamming.

4. The Classical Boltzmann Equation

The notion of entropy is important for thermal systems because it satisfies the second law,

$$\frac{\partial S}{\partial t} \geq 0, \tag{10}$$

which states that there is a maximum entropy state which, according to the evolution in Eq. (10), any system evolves toward, and reaches at equilibrium. A semi-rigorous proof of the Second Law was provided by Boltzmann (the well-known 'H-theorem'), by making use of the 'classical Boltzmann equation', as it is now known.

In order to derive this equation, Boltzmann made a number of assumptions concerning the interactions of particles. The most important of these assumptions were:

- The collision processes are dominated by two-body collisions (Fig. 3a). This is a plausible assumption for a dilute gas since the system is of very low density, and the probability of there being three or more particles colliding is infinitesimal.
- Collision processes are uncorrelated, i.e. all memory of the collision is lost on completion and is not remembered in subsequent collisions: the famous Stosszahlansatz. This is also valid only for dilute gases, but the proof is more subtle.

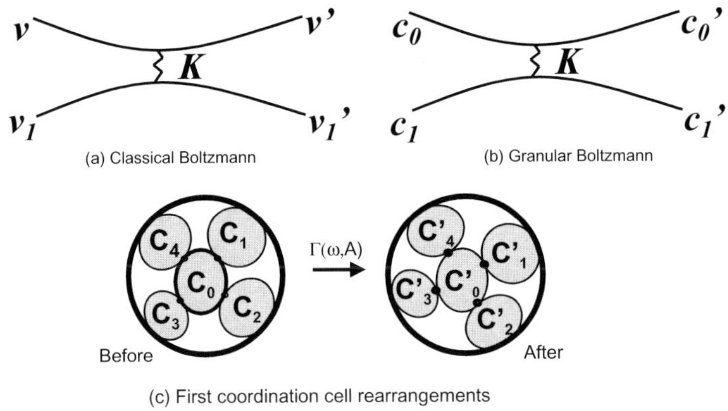

Figure 3. (a) Collision of two particles in a dilute gas. (b) "Collision of two configurations" given in terms of two contact points in a jammed material. (c) Rearrangements inside a pocket a under the first coordination shell approximation of grain $\alpha = 0$.

Thus, Boltzmann proves Eq. (10) for a dilute gas only, but this is a readily available situation. The remaining assumptions have to do with the kinematics of particle collisions, i.e. conservation of kinetic energy, conservation of momentum, and certain symmetry of the particle scattering cross-sections.

Let $f(v,r)$ denote the probability of a particle having a velocity v at position r. This probability changes in time by virtue of the collisions. The two particle collision is visualised in Fig. 3a where v and v_1 are the velocities of the particles before the collision and v' and v_1' after the collision.

On time scales larger than the collision time, momentum and kinetic energy conservation apply:

$$mv + mv_1 = mv' + mv_1', \qquad \frac{1}{2}mv^2 + \frac{1}{2}mv_1^2 = \frac{1}{2}mv'^2 + \frac{1}{2}mv_1'^2. \tag{11}$$

Then, the distribution $f(v,r)$ evolves with time according to

$$\frac{\partial f}{\partial t} + v\frac{\partial f}{\partial r} + \int \mathcal{K}(v, v'; v_1, v_1') \ \left(f(v)f(v_1) - f(v')f(v_1')\right) \ d^3v_1 d^3v' d^3v_1' = 0. \tag{12}$$

The kernel $\mathcal{K}$ is positive definite and contains δ-functions to satisfy the conditions (11), the flux of particles into the collision and the differential scattering cross-section. We consider the case of homogeneous systems, i.e. $f = f(v)$, and define

$$S = -k_B \int f \log f. \tag{13}$$

Defining $x = ff_1/f'f'_1$ we obtain

$$\frac{\partial S}{\partial t} = \int \mathcal{K} \, \log x \, (1-x) d^3v_1 \, d^3v' \, d^3v'_1, \qquad (1-x)\log x \geq 0, \qquad \mathcal{K} \geq 0. \tag{14}$$

Hence $\partial S/\partial t \geq 0$ (see standard text books on statistical mechanics).

It is also straightforward to establish the equilibrium distribution where $\partial S/\partial t = 0$ since it occurs when the kernel term vanishes. This occurs when the condition of detailed balance is achieved, $x = 1$:

$$f(v)f(v_1) = f(v')f(v'_1). \tag{15}$$

The solution of Eq. (15) subjected to the condition of kinetic energy conservation is given by the Boltzmann distribution

$$f(v) = \left(\frac{k_B T}{m\pi}\right)^{3/2} e^{-\frac{1}{2}\beta m v^2}, \tag{16}$$

where $\beta = 1/k_B T$. Equation (16) is a reduced distribution and valid only for a dilute gas. The Gibbs distribution represents the full distribution and is obtained by replacing the kinetic energy in (16) by the total energy of the state to obtain:

$$P(E) \sim e^{-\beta E}. \tag{17}$$

The question is whether a similar form can be obtained in a granular system in which we expect

$$P(\mathcal{W}) \sim e^{-\mathcal{W}/\lambda X}, \tag{18}$$

where X is the compactivity in analogy with $T = \partial E/\partial S$. Such an analysis is shown in the next section in an approximate manner.

5. 'Boltzmann Approach' to Granular Matter

The analogous approach to granular materials consists in the following: the creation of an ergodic grain pile suitable for a statistical mechanics approach via a tapping method for the exploration of the available configurations analogous to Brownian motion, the definition of the discrete elements tiling the granular system via the volume function $\mathcal{W}$, and an equivalent argument for the energy conservation expressed in terms of the system volume necessary for the construction of the Boltzmann equation.

We have already established the necessity of preparing a granular system adequate for real statistical mechanics so as to emulate ergodic conditions. The grain motion must be well-controlled, as the configurations available to the system will be dependent upon the amount of energy/power put into the system. This pretreatment is analogous to the averaging which takes place inherently in a thermal system and is governed by temperature.

As explained, the granular system explores the configurational landscape by the external tapping introduced by the experimentalist. The tapping is characterised by a

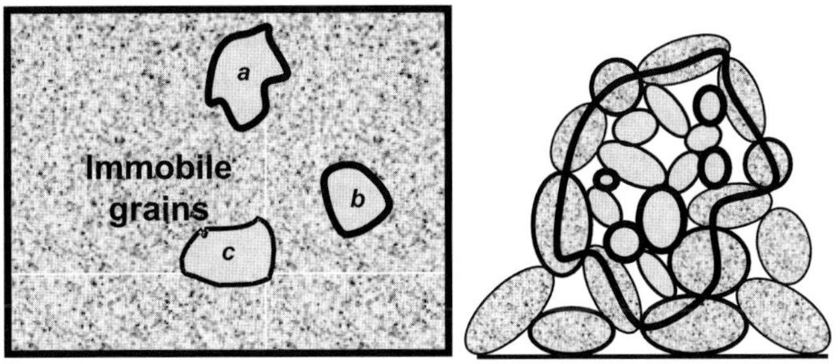

Figure 4. (a) Regions of mobile grains a, b, c in a matrix of immobile grains below the Coulomb threshold. (b) Detail of pocket of mobile grains a surrounded by immobile grains which are shaded.

frequency and an amplitude (ω, Γ) which cause changes in the contact network, according to the strength of the tap. The magnitude of the forces between particles in mechanical equilibrium and their confinement determine whether each particle will move or not. The criterion of whether a particular grain in the pile will move in response to the perturbation will be the Mohr-Coulomb condition of a threshold force, above which sliding of contacts can occur and below which there can be no changes. The determination of this threshold involves many parameters, but it suffices to say that a rearrangement will occur between those grains in the pile whose configuration and neighbours produce a force which is overcome by the external disturbance.

The concept of a threshold force necessary to move the particles implies that there are regions in the sample in which the contact network changes and those which are unperturbed, shown in Fig. 4. Of course, since this is a description of a collective motion behaviour, the region which can move may expand or contract, but the picture at any moment in time will contain pockets of motion encircled by a static matrix. Each of these pockets has a perimeter, defined by the immobile grains. It is then possible to consider the configuration before and after the disturbance inside this well-defined geometry.

The present derivation assumes the existence of these regions. It is equivalent to the assumption of a dilute gas in the classical Boltzmann equation, although the latter is readily achieved experimentally.

The energy input must be on the level of noise, such that the grains largely remain in contact with one another, but are able to explore the energy landscape over a long period of time. In the case of external vibrations, the appropriate frequency and amplitude can be determined experimentally for different grain types, by investigating the motion of the individual grains or by monitoring the changes in the overall volume fraction over time. It is important that the amplitude does not exceed the gravitational force, or else the grains are free to fly up in the air, re-introducing the problem of initial creation just as they would if they were simply poured into another container.

Within a region a we have a volume $\sum_{\alpha \in a} \mathcal{W}^{\alpha}$ and after the disturbance a volume which is now $\sum_{\alpha \in a} \mathcal{W}'^{\alpha}$ as seen in Fig. 3c. In Section 3.1 we have discussed how to define the volume function $\mathcal{W}^\alpha$ as a function of the contact network. Here the simplest "one grain" approximation is used as the "Hamiltonian" of the volume as defined by Eq. (6). In reality it is much more complicated, and although there is only one label α on the contribution of grain α to the volume, the characteristics of its neighbours may also appear. Instead of energy being conserved, it is the total volume which is conserved while the internal rearrangements take place within the pockets described above. Hence

$$\sum_{\alpha\in a} \mathcal{W}^\alpha = \sum_{\alpha\in a} \mathcal{W}'^\alpha \tag{19}$$

We now construct a Boltzmann equation. Suppose z particles are in contact with grain $\alpha = 0$, as seen in Fig. 3c. For rough particles $z = 4$ while for smooth $z = 6$ at the isostatic limit. The probability distribution will be of the contact points which are represented by the tensor $\mathcal{C}^\alpha$, Eq. (5), for each grain, where α ranges from 0 to 4 in this case. So the analogy of $f(v)$ for the Boltzmann gas equation becomes $f(\mathcal{C}^0)$ for the granular system and represents the probability that the external disturbance causes a particular motion of the grain. We therefore wish to derive an equation

$$\frac{\partial f(\mathcal{C}^0)}{\partial t} + \int \mathcal{K}(\mathcal{C}^\alpha,\mathcal{C}'^\alpha)\ \left(f_0 f_1 f_2 f_3 f_4 - f'_0 f'_1 f'_2 f'_3 f'_4\right)\ d\mathcal{C}'^0 \prod_{\alpha\neq 0} d\mathcal{C}^\alpha d\mathcal{C}'^\alpha = 0 \tag{20}$$

The term $\mathcal{K}$ contains the condition that the volume is conserved (19), i.e. it must contain $\delta(\sum \mathcal{W}^\alpha - \sum \mathcal{W}'^\alpha)$. The cross-section is now the compatibility of the changes in the contacts, i.e. $\mathcal{C}^\alpha$ must be replaced in a rearrangement by $\mathcal{C}'^\alpha$ (unless these grains part and make new contacts in which case a more complex analysis is called for). We therefore argue that the simplest $\mathcal{K}$ will depend on the external disturbance Γ, ω and on $\mathcal{C}^\alpha$ and $\mathcal{C}'^\alpha$, i.e.

$$\frac{\partial f(\mathcal{C}^0)}{\partial t} + \int \delta\Big(\sum_\alpha \mathcal{W}^\alpha - \sum_\alpha \mathcal{W}'^\alpha\Big)\ \mathcal{J}(\mathcal{C}^\alpha,\mathcal{C}'^\alpha)\ \Big(\prod_{\alpha=0}^{z} f_\alpha - \prod_{\alpha=0}^{z} f'_\alpha\Big) d\mathcal{C}'^0 \prod_{\alpha\neq 0} d\mathcal{C}^\alpha d\mathcal{C}'^\alpha = 0 \tag{21}$$

where $\mathcal{J}$ is the cross-section and it is positive definite.

The Boltzmann argument now follows. As before

$$S = -\lambda \int f \log f \qquad\qquad x = \frac{f_0 f_1 f_2 f_3 f_4}{f'_0 f'_1 f'_2 f'_3 f'_4}, \tag{22}$$

and

$$\frac{\partial S}{\partial t} \geq 0, \tag{23}$$

the equality sign being achieved when $x = 1$ and

$$f_\alpha = \frac{e^{-\mathcal{W}^\alpha/\lambda X}}{Z}, \tag{24}$$

with the partition function

$$Z = \sum e^{-\mathcal{W}^\alpha/\lambda X}\quad \Theta, \tag{25}$$

and the analogue to the free energy being $Y = -X \ln Z$, and $X = \partial V / \partial S$.

The detailed description of the kernel $\mathcal{K}$ has not been derived as yet due to its complexity. Just as Boltzmann's proof does not depend on the differential scattering cross section, only on the conservation of energy, in the granular problem we consider the steady state excitation externally which conserves volume, leading to the granular distribution function, Eq. (24).

It is interesting to note that there is a vast and successful literature of equilibrium statistical mechanics based on $\exp(-\mathcal{H}/k_B T)$, but a meagre literature on dynamics based on attempts to generalise the Boltzmann equation or, indeed, even to solve the Boltzmann equation in situations remote from equilibrium where it is still completely valid. It means that any advancement in understanding how it applies to analogous situations is a step forward.

5.1. Experimental Validation of the Statistical Mechanics Concepts

The first step in realising the idea of a general statistical jamming theory is to understand in detail the characteristics of a jammed configuration in particulate systems. Next we present a novel experimental method to explore this problem using confocal microscopy [12].

The key feature of this optical microscopy technique is that only light from the focal plane is detected. Thus 3D images of translucent samples can be acquired by moving the sample through the focal plane of the objective and acquiring a sequence of 2D images. Our model system consists of a dense packing of emulsion oil droplets, with a sufficiently elastic surfactant stabilising layer to mimic solid particle behaviour, suspended in a continuous phase fluid. The refractive index matching of the two phases, necessary for 3D imaging, is not a trivial task since it involves unfavourable additions to the water phase, disturbing surfactant activity. The successful emulsion system, stable to coalescence and Ostwald ripening, consisted of Silicone oil in a solution of water ($w_t = 50\%$) and glycerol ($w_t = 50\%$), stabilised by $0.01mM$ sodium dodecylsulphate (SDS). The droplet phase is fluorescently dyed using Nile Red, prior to emulsification. The control of the particle size distribution, prior to imaging, is achieved by applying very high shear rates to the sample, inducing droplet break-up down to a radius mean size of $2\mu m$. Since the emulsion components have different densities, the droplets cream under gravity to form a random close packed structure. In addition, the absence of friction ensures that the system has no memory effects and reaches a true jammed state before measurement.

The 3D reconstruction of the 2D slices is shown in Fig. 5. We have developed a sophisticated image analysis algorithm which uses Fourier Filtering to determine the particle centres with subvoxel accuracy. Previously, we developed a method to measure the interdroplet forces and their distribution in the sample volume [12]. Using an extension to the same image analysis method, the 3D images of a densely packed particulate model system now allow for the characterisation of the volume function $\mathcal{W}$, by the partitioning of the images into first coordination shells of each particle, described in Section 3.1. The polyhedron obtained by such a partitioning is shown in Fig. 6.

We are able to test how the approximation in Eq. (6) for the volume compares with the actual volume measured from the image for each grain. Their correlation is shown in Fig. 7. This approximation works well for coordination numbers larger than 3 in 2D and even

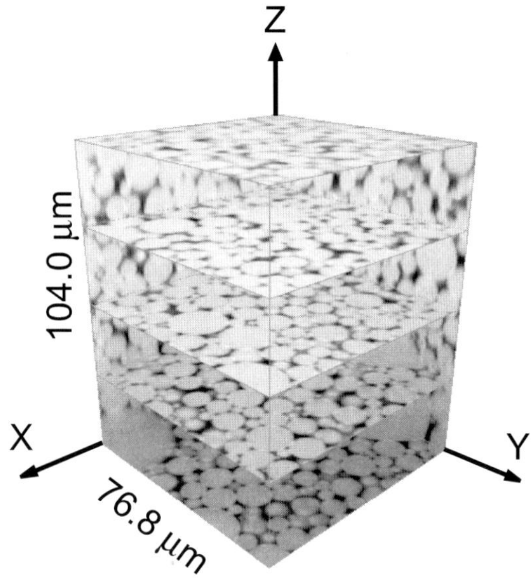

Figure 5. Confocal image of the densely packed emulsion system.

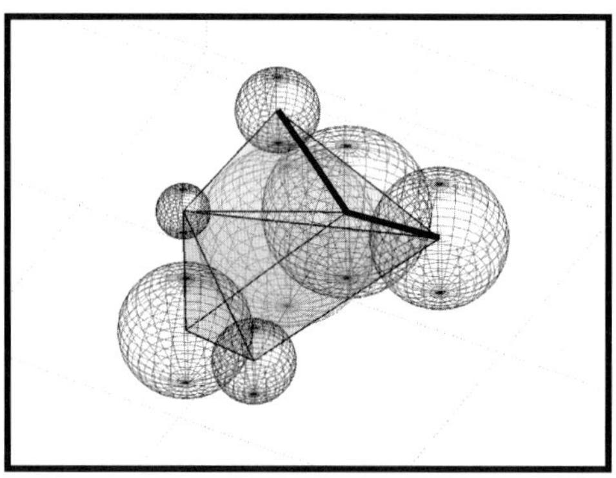

Figure 6. An example of a volume $\mathcal{W}$ as the polyhedron constructed from the 3D images. The centre grain (red) has 6 grains in contact (black), the centres of which are joined to form the polyhedron.

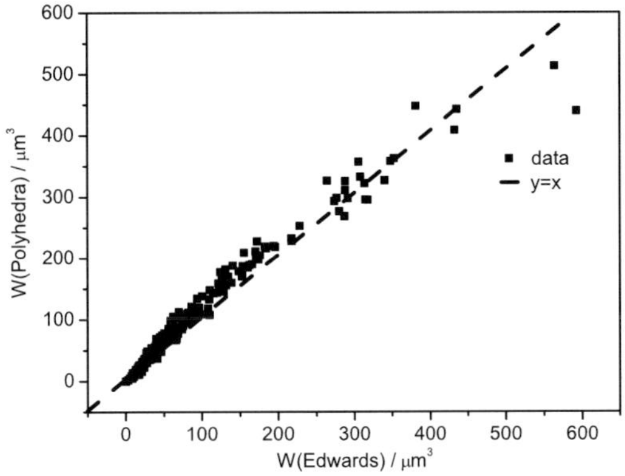

Figure 7. $\mathcal{W}$ function obtained from the configuration tensor $\mathcal{C}$ are plotted against the polyhedra constructed from the images.

in 3D, due to the partitioning of the obtained volumetric objects into triangles/pyramids, intrinsic to the method, and subsequently summing over them to obtain the resulting volume. It is clear that very large volumes, belonging to grains with high coordination numbers, do stray from the theoretical value due to the complex geometries involved. According to the experimental measurements of $\mathcal{W}$ employed in this study, the total volume of the system was found to be overestimated by only $\approx 5\%$.

The ability to measure this function and therefore its fluctuations in a given particle ensemble, enables the calculations of the macroscopic variables. In Fig. 8 we show the probability distribution of $\mathcal{W}$ showing an exponential behaviour as given by Eq. (24). The exponential probability distribution of $\mathcal{W}$ leads to the compactivity X according to Eq. (18). This implies that we can arrive at the thermodynamic system properties from the knowledge of the microstructure. Many images, i.e. configurations, can be treated in this way to test whether system size influences the macroscopic observables. If the particles are subjected to ultracentrifugation resulting in configurations of a higher density, the influence of pressure on the macroscopic variables can also be tested.

Such a characterisation of the governing macroscopic variables, arising from the information of the microstructure, allows one to predict the system's behaviour through an equation of state. This is the first experimental study of such statistical concepts in particulate matter and opens new possibilities for testing the above described thermodynamic formulation. In principle, one can apply low amplitude vibrations to the system and observe the droplet configuration before and after the perturbation, thus testing the ideas proposed in the Boltzmann derivation.

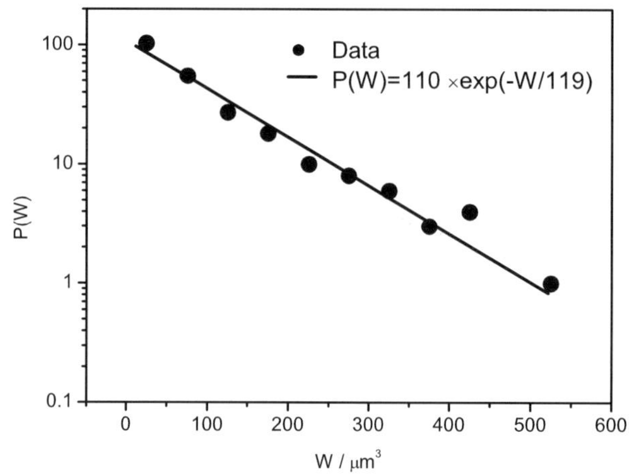

Figure 8. Probability distribution of $\mathcal{W}$ fitted with a single exponential. The decay constant is the compactivity, $X = 119\mu m^3/\lambda$.

Acknowledgments. We thank D. Grinev and R. Blumenfeld for stimulating discussions. H. Makse acknowledges financial support from the National Science Foundation, DMR-0239504 and the Department of Energy, Division of Materials Sciences and Engineering, DE-FE02-03ER46089.

REFERENCES

1. E. R. Nowak, J. B. Knight, E. BenNaim, H. M. Jaeger and S. R. Nagel. *Phys. Rev. E* **57**, 1971 (1998).
2. S. F. Edwards and R. B. S. Oakeshott, *Physica A* **157**, 1080-1090 (1989).
3. S. F. Edwards, in *Granular matter: an interdisciplinary approach* (ed. A. Mehta) 121-140 (Springer-Verlag, New York, 1994).
4. H. A. Makse, J. Brujić, and S. F. Edwards, in *The Physics of Granular Media*, (eds. H. Hinrichsen and D. E. Wolf) (Wiley-VCH, 2004).
5. R. C. Ball and R. Blumenfeld, *Phys. Rev. Lett.* **88**, 115505 (2002).
6. P. Philippe, and D. Bideau, *Europhys. Lett.* **60**, 677 (2002).
7. J. Brujić, D. L. Johnson, O. Sindt, and H. A. Makse (Schlumberger report, to be published).
8. A. Chakravarty, S. F. Edwards, D. V. Grinev, M. Mann, T. E. Phillipson, A. J. Walton, Proceedings of the Workshop on Quasi-static Deformations of Particulate Materials, to be published.
9. H. A. Makse and J. Kurchan, *Nature* **415**, 614-617 (2002).
10. S. B. Savage, *Adv. Appl. Mech.* **24**, 289-365 (1994).

11. J. T. Jenkins, and S. B. Savage, *J. Fluid Mech.* **130**, 187-202 (1983).
12. J. Brujić, S. F. Edwards, I. Hopkinson, and H. A. Makse, *Physica* A **327**, 201-212 (2003).

Unifying Concepts in Granular Media and Glasses
A. Coniglio, A. Fierro, H.J. Herrmann and M. Nicodemi (editors)

A possible experimental test of the thermodynamic approach to granular media

D.S. Dean[a] and A. Lefevre[a]

[a]Laboratoire de Physique Theorique, IRSAMC Université Paul Sabatier, 118 route de Narbonne, 31062 Toulouse Cedex 04, France.

Some tests of the thermodynamic approach to describe the steady state of driven granular media are discussed. The application of the canonical form of the Edwards measure to realistic granular media is extremely difficult as the entropy of blocked states is not easily calculated. We propose a test of a weak form of the Edwards measure based on comparing the fluctuations in the system at different driving strengths. To illustrate the tests proposed we study the steady state distribution of the energy of the Sherrington Kirkpatrick model driven by a tapping mechanism which mimics the mechanically driven dynamics of granular media. The dynamics consists of two phases: a zero temperature relaxation phase which leads the system to a metastable state then a tapping which excites the system thus reactivating the relaxational dynamics. Numerically we investigate whether the distribution of the energies of the blocked states obtained agrees with a simple canonical form of the Edwards measure. It is found that this canonical measure is in good agreement with the dynamically measured energy distribution.

1. Introduction

Complex systems such as granular media possess a large number of metastable or blocked configurations. When a granular medium is shaken it quickly relaxes into a blocked configuration, a subsequent shake of tap will lead it to another blocked or jammed state and so on. If the driving mechanism is held constant one expects the system to enter into a quasi-equilibrium stationary state. Various driving mechanisms can be investigated experimentally such, as vertical tapping [1] and horizontal shaking [2]. In granular media and other complex systems such as spin glasses, the entropy of these blocked states is extensive in the system size, and hence it has been proposed that one may use a thermodynamic measure over blocked states to describe this steady state. The simplest proposition is that the system is characterised by a number of quantities which are fixed on average and then the measure on the steady state is obtained from the maximum entropy state (on blocked states) with the relevant macroscopic quantities fixed [3]. This simple idea has recently been investigated in a wide range of systems and has been shown to be relatively successful. Various tests of the applicability of these thermodynamic ideas have been carried out and although some confirmation has been made in more realistic sheared granular systems [4] most work has been carried out on simpler model systems which one hopes capture the basic physics of granular media. The Edwards flat measure has been shown to be of predictive value in some simple one dimensional lattice models [5], however there are clearly examples where the approach fails [6], however recently it was shown that more sophisticated versions of the Edwards measure introducing by ensembles with several quantities fixed on average can remedy the deficiencies of the basic measure in these cases [7]. Other toy models that have been analysed are lattice based models with kinetic constraints in higher dimensions [8,9] and also spin glass models [10] where non-thermal driving is used to move the system between blocked states. One point in common with many of the tests of the Edwards measure is the conclusion that it works best at low driving rates.

Even if it is not expected to be exact many systems may be described to a good engineering level by these measures. Given the difficulty of the analysis of the highly nonlocal dynamics in these systems this is an important step toward understanding their steady state regimes. There is no clear ergodicity in these systems and no detailed balance as in usual statistical mechanics. Edwards argued that a system might conceivably explore blocked configurations in a flat manner if the driving involved extensive manipulations meaning the displacement of a macroscopic number of particles, for example shaking, stirring or pouring granular media. An interesting consequence of the applicability of thermodynamic ideas is that one may describe phase transitions in these driven systems [11]. Solving the dynamics of driven dense granular media thus seems technically formidable and a rigorous derivation of the steady state distribution in such systems also seems out of reach. To date, at best, the Edwards measure can thus be regarded as an ansatz for the steady state distribution of an evolution equation we don't know how to write down ! Despite the lack of a rigorous foundation the approach seems worth following up as, where it does work, it gives an accurate description of phenomenon where, due to the nonlocal nature of the dynamics, we had no theory at all.

As mentioned above a good theoretical and numerical testing ground for this thermodynamic approach to granular media are spin glasses. Spin glasses have an exponentially large (with the system size) number of metastable states and hence an extensive entropy of blocked states as do granular media. The definition of a blocked state in a spin glass simulated on a computer depends of course on the local dynamics. Under single spin flip dynamics a metastable state is one where flipping any single spin increases the energy, it is thus a blocked state under any single spin flip Monte Carlo dynamics. Various spin glass models have been studied to explore the accuracy of the Edwards measure as a function of the relaxational dynamics and the tapping mechanism [10]. Here we shall explore the driven dynamics of the Sherrington-Kirkpatrick (SK) [12] spin glass model defined by the Hamiltonian

$$H = -\sum_{i<j} J_{ij} S_i S_j \tag{1}$$

where the S_i ($1 \leq i \leq N$) are Ising spins and the interaction J_{ij} are independent Gaussian random variables of zero mean and variance $\overline{J_{ij}^2} = 1/N$. A blocked or metastable state for single spin flip dynamics in this model is one where $S_i = \text{sign}(\sum_{j \neq i} J_{ij} S_j)$ for all i, *i.e* every spin is aligned with its local field.

We investigate whether at fixed tapping rate (to be defined later) the states explored dynamically obey a form of Boltzmann distribution. The applicability of a Boltzmann distribution, which is a consequence of the canonical form of the Edwards measure, for tapped spin glass systems seems not to have been investigated to date. The results presented here are quite striking, despite a lack of detailed balance we shall see that a Boltzmann distribution describes excellently the histogram of the energies of the blocked states during tapping.

The tapping mechanism introduced in spin glass models involves two steps: 1) Relaxation: the system evolves under zero temperature Glauber dynamics until it is blocked in a metastable state; 2) The system is tapped with tapping parameter p, that is to say each spin is flipped in parallel with probability p - this is the extensive manipulation as at each tap on average p spins are flipped. Given a tapping rate p, it is observed that the system reaches a steady state regime with a fixed average energy per spin ϵ. If we postulate the canonical form of the Edwards measure for the quasi-equilibrium distribution of the driven dynamics (observed on the blocked states) one has the steady state distribution of energy given by

$$\rho_{EDW}(E) = \frac{N_{MS}(E)\exp(-\beta E)}{Z} \tag{2}$$

where $N_{MS}(E)$ is the number of metastable states of energy E, β is the inverse Edwards temperature which can be thought of as a Lagrange multiplier fixing the average energy per spin and Z is the canonical partition function $Z = \sum_E N_{MS}(E)\exp(-\beta E)$. We note that a fully flat measure with fixed energy per spin would be $P_\alpha = \exp(-\beta E_\alpha)/Z$, where E_α is the energy of the blocked state α. In this paper we shall investigate only the weaker form of the measure Eq. (2).

The hypothesis Eq. (2) will be tested in the following manner. We fix a bin size for the energies ΔE and by exact enumeration we compute $N_{MS}(E)$ for a given system and realisation of the disorder. The exact enumeration means that we can explore system sizes up to $N = 30$. Then the tapping dynamics is applied to the same system and the dynamical histogram $\rho_{DYN}(E)$ of the energy with the same bin size ΔE is computed over 10^6 taps after the system has been tapped 10^6 times to reach the steady state. We then calculate

$$R(E) = \frac{\rho_{EDW}(E)}{N_{MS}(E)} \tag{3}$$

If the Edwards measure in the form Eq. (2) is valid then one should obtain

$$r(E) = \ln(R(E)) = -\beta E + C \tag{4}$$

where C is a constant which should be independent of E. If Eq. (2) is valid then β should also be independent of E. If Eq. (4) is valid then another interesting question to as is what is the dependence of β on a) the tapping strength p, b) the size of the system N c) the realisation of the disorder. It is interesting to see if the tapping strength p determines a unique temperature in the limit of large system sizes, for systems with the same statistical distribution of couplings *i.e.* is β self averaging ? We shall see that the results obtained here for small system sizes suggest that Eq. (4) is obeyed on increasing the system size and that the parameter β becomes independent of the system size and the disorder realisation for large systems.

In order to see what happens for larger systems (where the enumeration of metastable states is no longer feasible), where the thermodynamic approach should be expected to work better, one may compare the dynamical distribution $N_D(E)$ measured for the steady states at different tapping rates. If the Edwards measure holds ,*i.e.* $N_D(E) \propto \rho_{EDW}(E)$, and if β is a self averaging quantity depending only on p, then for any two tapping strengths p and p' we should find that

$$\frac{N_D(E,p)}{N_D(E,p')} = \exp\left(-(\beta(p) - \beta(p'))E\right) \tag{5}$$

and hence

$$\begin{aligned} c(p,p',\epsilon) &= \frac{\ln(N_D(E,p)) - \ln(N_D(E,p'))}{N} \\ &= -(\beta(p) - \beta(p'))\epsilon \end{aligned} \tag{6}$$

where $\epsilon = E/N$ is the energy per spin. This a would constitute a weak verification of the Edwards hypothesis but has the merit of being practicable for large system sizes. This particular test is the most useful for realistic purposes where $N_{MS}(E)$ can neither be calculated analytically or evaluated by direct counting. The main message of this paper is not that the Edwards measure passes the tests proposed here on the SK model reasonably well but that the test above embodied by Eq. (6) could be applied to a whole range of system with steady state regimes to see if the statistics of the steady state regime are described by a Boltzmann re-weighting of some structure function $N_S(E)$ *i.e.*

$$N_D(E,p) = N_S(E)\exp\left(-\beta(p)E\right) \tag{7}$$

It is of course possible (and even probable if we believe that the Edwards measure should not work generically) that $N_S(E)$ depends on the type of driving mechanism and on the precise form of the single spin flip relaxational dynamics.

A final test in the context of the tapped SK model is to use the analytic, but annealed calculation, of the average number of metastable states at fixed energy, $\overline{N_{MS}(\epsilon)}$ (the overline indicates the disorder average over realisations), which gives in the thermodynamic limit

$$\overline{N_{MS}(\epsilon)} \approx \exp(N S^a_{MS}(\epsilon)) \tag{8}$$

where [13,14]

$$S^a_{MS}(\epsilon) = \min_z \left\{ \frac{1}{2}z^2 - E^2 + \ln\left(1 - \operatorname{erf}\left(\frac{z+2E}{\sqrt{2}}\right)\right)\right\} \tag{9}$$

is the annealed entropy per spin of metastable states of energy ϵ per spin. If the Edwards measure holds then one should find

$$\frac{\ln\left(N_D(\epsilon,p)\right) - \ln\left(N_{MS}(\epsilon)\right)}{N} = -\beta(p)\epsilon + C \tag{10}$$

We further expect that the dynamical entropy per spin $S_D(\epsilon,p) = \ln(N_D(\epsilon,p))/N$ and the entropy of metastable states per spin $S_{MS}(\epsilon) = \ln(N_{MS}(\epsilon,p))/N$ are self averaging and thus Eq. (10) may be written for large N as

$$r(\epsilon) = \overline{S_D(\epsilon,p)} - \overline{S_{MS}(\epsilon)} = -\beta(p)\epsilon + C \tag{11}$$

Unfortunately no exact calculation for $\overline{S_{MS}(\epsilon)}$ exists, however it is known that $S^a_{MS}(\epsilon) = \overline{S_{MS}(\epsilon)}$ for $\epsilon_c > -0.672$ [14]. Furthermore this annealed approximation seems to be good, though not exact, to even lower energies as confirmed by a replica symmetric calculation [15]. Hence at energies above ϵ_c we have the formula

$$r_a(\epsilon) = \overline{S_D(\epsilon, p)} - S^a_{MS}(\epsilon) = -\beta(p)\epsilon + C \tag{12}$$

which should also work well approximately at energies below, but not too far from, ϵ_c. We remark here that by Jensen's inequality $S^a_{MS}(\epsilon) \geq S_{MS}(\epsilon)$.

We now discuss the results of these numerical tests. Systems between size 15 and 30 were analysed. The numerically measured value of $r(E)$ was computed from the static metastable state energy histogram and the histogram obtained by tapping the system 10^6 times. This was done for values of p, 0.1, 0.2, 0.3, 0.4,and 0.5. Increasing the number of taps did not change the dynamical histogram so we can be sure that we are measuring the steady state (this is natural given the small system sizes). It is possible that with very low tapping rates the time to reach the steady state even for a system of 30 spins becomes large. For example the results for $p = 0.3$ shown on Fig. (1) we see that for small systems sizes the points of the numerically measured $R(E)$ are scattered about the straight line fit predicted by Eq. (4). However as the system size is increased the straight line fit appears excellent. The deviations from the straight line fit also appear to be non systematic. Similarly excellent fits are also obtained for all the tapping rates studied. In the case of the SK model, a Boltzmann distribution of the energies of the metastable states visited in the steady state regime appear to depend simply on having a large enough system size. It is generally believed that the Edwards measure should be more accurate in gently driven systems, here the weak form characterised by Eq. (4) seems equally valid at all tapping rates. We also note that the slope of the straight lines appear to be saturating to a constant value of $\beta(p)$ on increasing N, suggesting that in the thermodynamic limit the Edwards temperature is fully characterised by the tapping rate p.

From this first numerical study one may conclude that if the weak form of the Edwards measure holds then it relies on the system being sufficiently large. The results above however seem encouraging, especially when one takes into account that average total number of metastable states in a system of size $N \sim \exp(0.1992N)$ [13] which is about 400 for a system of size 30

We now carry out simulations for a system of size 200, here the average number of metastable states is about 2×10^{17}. The bin size is also refined to $0.0025N$ Naively one would expect the thermodynamics approach to work better in this limit as the phase space of metastable states is much bigger. Also there are many more states in each energy bin and hence it is really the weak form of the Edwards measure that is tested. Shown on Fig. (2) is the numerically computed $c(p, p,' \epsilon)$ from a simulation of the same system tapped at values of p, 0.1, 2., 0.3, 0.4 and 0.5. The system was first equilibrated by tapping 5000 times and the dynamical histogram was constructed over 5×10^6 taps. One sees from Fig. (2) that the values of $c(p, p', \epsilon)$ are to a very high degree of accuracy on a straight line as predicted by Eq. (6), the deviations at low energies, where the sampling is lowest, appear non-systematic. Note that if the logarithm of the dynamical histograms $\ln(N_D(\epsilon, p)$, rather than the differences, is plotted one sees that in the region where the histograms overlap each of the $\ln(N_D(\epsilon, p)$ has a distinct curvature and thus one is not obtaining a straight line by subtracting two straight lines. As the system size is increased one will find that the range of ϵ visited in the histogram shrinks (as $1/\sqrt{N}$) about the mean value $\epsilon(p)$ at that tapping rate. In order to have a significant range of overlap between two different tapping rates therefore, one must have that the system size is not too large and that the tapping rates compared, p and p', are sufficiently close so that for the N considered the centres of the histograms $\epsilon(p)$ and $\epsilon(p')$ are close enough to have a significant overlap.

The energy histograms generated by the precedent simulations were used to compute $r_a(\epsilon)$, the results are shown in Fig. (3). The curves have been shifted vertically for clarity. A straight line fit was performed in the region $\epsilon > \epsilon_c$ and we see that in this region the fit is excellent. The lower energy part of the curves also appear linear down to around $\epsilon = -0.72$. The replica symmetric calculation of [15] also visibly (on the same scale) departs from the annealed calculation near this energy, although we emphasise that this is not an exact calculation but can be expected to be closer to the full quenched result than the annealed one. The the lower tapping rates explore the lower energy regime while the higher rates explore the higher energy regime. The deviations above ϵ_c from the annealed entropy are at the high energy end of the numerical histogram where the sampling is smallest. Also shown on Fig. (3) is $-S^a_{MS}(\epsilon)$ which has a very pronounced curvature showing that the fact that $r_a(\epsilon)$ is a straight-line is non-trivial.

To summarise we see that the assumption of an effective Boltzmann distribution for the energy of the states explored by tapping dynamics describes extremely well the numerically obtained results. This description seems to work better on increasing the system size for small systems and seems to be valid for larger systems via slightly more indirect tests. Why it works is rather mysterious and if it is exact it would be an extremely useful method to numerically map out the distribution of metastable states in various spin glass models. The relaxational dynamics used was random update which uses a lot of computer time for these simulations. Sequential update is much quicker for these simulations and we have carried out such simulations and found to within very small deviations the same results: If one could prove the applicability of the Edwards measure for these types of dynamics one would have an extremely powerful method to explore metastable states and inherent states in glassy systems. The standard method employing an auxiliary Hamiltonian [8] is much more time consuming - although classical statistical mechanics tells us that it will give the right result.

In the original proposition of Edwards in the context of granular media, the steady state volume fraction V occupied by a driven granular media is described in the canonical approach by

$$\rho_D(V) \propto \exp(-\frac{V}{X} + S(V)) \tag{13}$$

where X is the compactivity and $S(V)$ is the entropy of blocked states occupying volume V. The problem of calculating $S(V)$ in a real system seems formidable. However it should be possible to test the prediction of Eq. (6) experimentally by comparing the histograms of V for systems driven at different driving amplitudes. As explained above the histograms should be compared for driving rates that are sufficiently close to have a sizable overlap in the distributions of the specific volume and for system sizes small enough to have significant fluctuations about the mean specific volume. This would be a crucial test, though we emphasise again, not a demonstration of the validity of the Edwards measure.

REFERENCES

1. E.R. Nowak, J.B. Knight, E. Ben-Naim, H.M. Jaeger, and S.R. Nagel,*Phys. Rev. E* **57**, 1971 (1998).
2. M. Nicolas, P. Duru and O. Pouliquen in *Compaction of Soils, Granulates and Powders*, D. Kolymbas and W. Fellin (eds.), A. A. Balkema, Rotterdam (2000)
3. S.F. Edwards in *Granular Media: An Interdisciplinary Approach* ed. A. Mehta (Springer-Verlag, New York, 1994); S.F. Edwards and A. Mehta, *Journal de Physique* **50**, 2489, (1989)
4. H.A. Makse and J. Kurchan, *Nature* **415**, 614 (2002)
5. J.J. Brey, A. Prados and B. Sanchez-Rey *Physica A* **275**,. 310 (2000); A. Lefèvre and D.S. Dean, *J. Phys. A* **34** L213 (2001).
6. J. Berg, S. Franz and M. Sellitto, *Eur. Phys. J. B* **26**, 349 (2002).
7. A. Lefèvre, *J. Phys. A* **35**, 9037 (2002).
8. A. Barrat, J. Kurchan, V. Loreto and M. Sellito, *Phys.Rev. Lett.* **85**, 5034 (2000); A. Barrat, J. Kurchan, V. Loreto and M. Sellito, *Phys. Rev. E* **63**, 051301 (2001)
9. A. Coniglio and M. Nicodemi, *Physica A* **296**, 451 (2001); A. Coniglio, A. Fierro and M. Nicodemi, *Physica A***302**, 193 (2001); A. Fierro, M. Nicodemi and A. Coniglio, *Europhys. Lett.* **59**, 642 (2002).
10. D.S. Dean and A. Lefèvre, *Phys. Rev. Lett.* **86** 5639 (2001); J. Berg and A. Mehta, *Europhys. Lett.* **56**, 784 (2001); D.S. Dean and A. Lefèvre, *Phys. Rev. E* **64**, 046110 (2001); L. Berthier, L.F. Cugliandolo and J.L. Iguian, *Phys. Rev. E* **63**, 051302 (2001); J. Berg and A. Mehta, *Phys. Rev. E.* **65**, 031305 (2002).
11. S.F. Edwards and R.B.S. Oakeshott, *Physica A* **157**, 1080 (1989); R.B.S. Oakeshott and S.F. Edwards, *Physica A* **202**, 482 (1994); A. Mehta and S.F. Edwards, *Physica A* **157**, 1091 (1989); A. Lefèvre and D.S. Dean, *Phys. Rev. B* **65**, 220403, (2002).
12. D. Sherrington and S. Kirkpatrick, *Phys. Rev. Lett.* **35**, 1792 (1975).
13. F. Tanaka and S.F. Edwards, *J. Phys. F.* **13** 2769 (1980).
14. A. J. Bray and M. A. Moore, *J. Phys. C* **13** L469. (1980).
15. S.A. Roberts *J. Phys. C.* **14**, 3015 (1981).

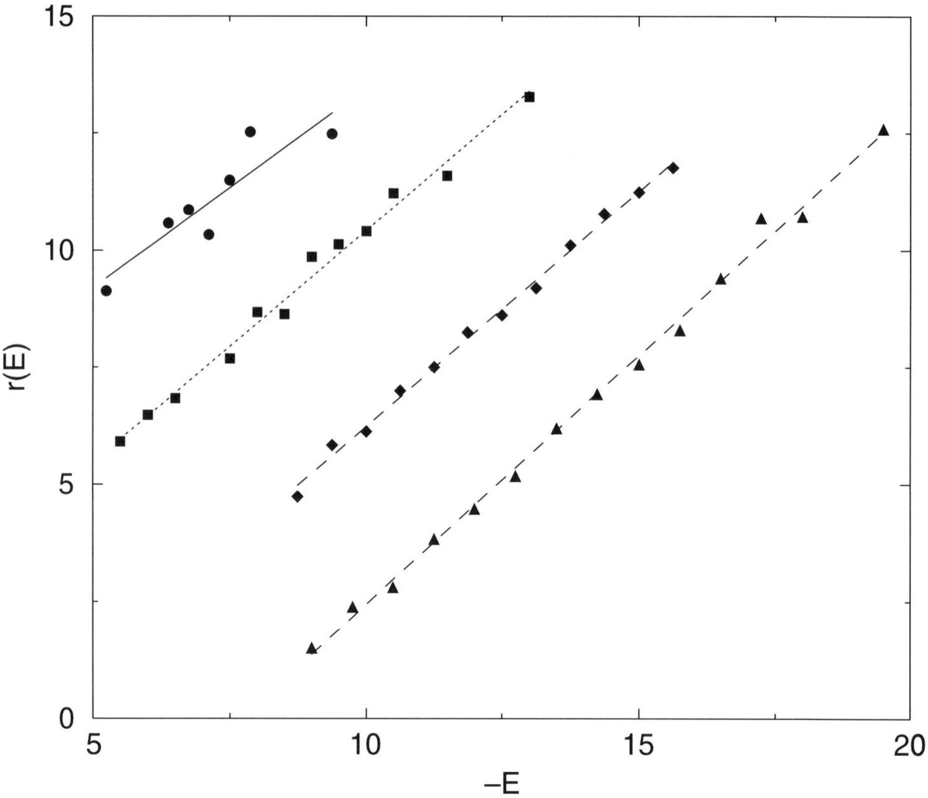

Figure 1. The computed values of $r(E)$ for the randomised spin flip relaxation at $p = 0.3$ with a bin size of $0.025N$ for systems sizes N of 15 (circles), 20 (squares), 25 (diamonds) and 30 (triangles). The straight lines are a linear fit to guide the eye.

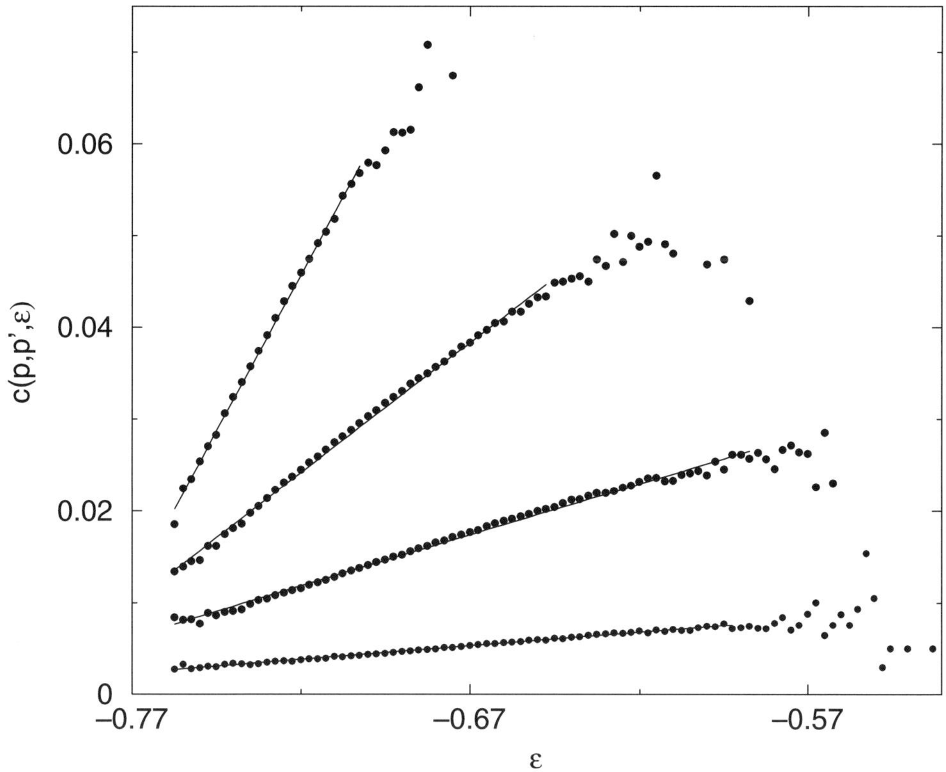

Figure 2. The numerically computed values of $c(p, p'\epsilon)$ from simulations of a system of size $N = 200$ spins. The curves have been vertically shifted for clarity. Shown in descending order are the results for (p, p'): (0.2,0.1), (0.3,0.2), (0.4,0.3), (0.4,0.5).

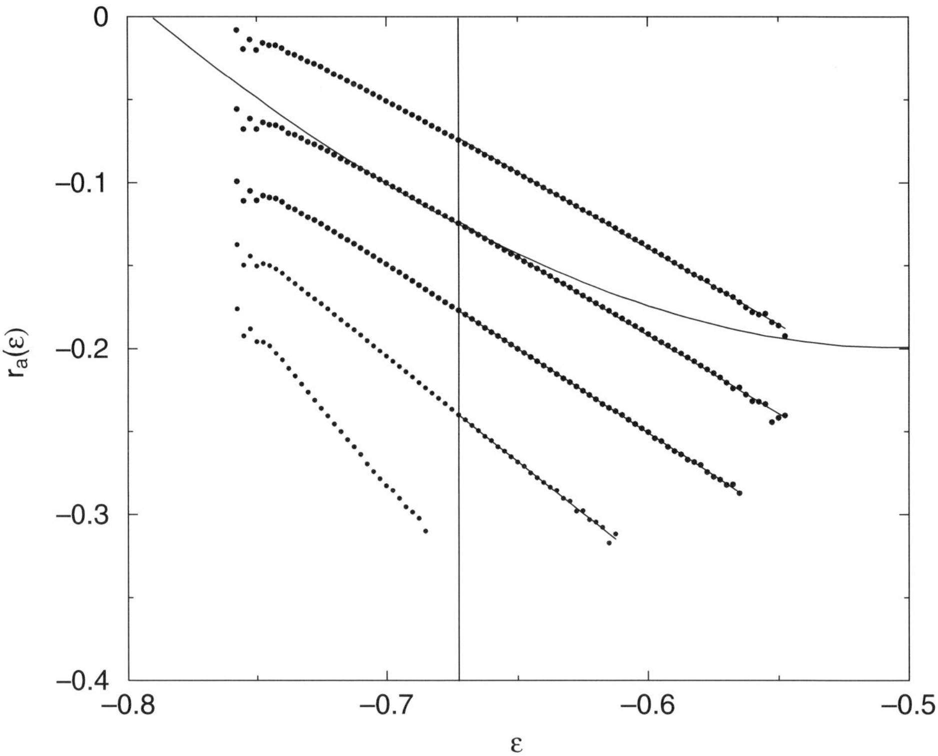

Figure 3. The calculation of $r_a(\epsilon)$ for a system with $N = 200$ and data taken over 5×10^6 taps. The straight line fits are shown in the region $\epsilon > \epsilon_c$ (to the right of the vertical line). The tapping rates are in ascending order $p =$ 0.1, 0.2, 0.3, 0.4 and 0.5. The solid line is $-S^a_{MS}(\epsilon)$ which has a clear curvature over the range of the numerically generated histograms

Unifying Concepts in Granular Media and Glasses
A. Coniglio, A. Fierro, H.J. Herrmann and M. Nicodemi (editors)

Memory and Kovacs effects in the parking-lot model: an approximate statistical-mechanical treatment

G. Tarjus[a] and P. Viot[a]

[a]Laboratoire de Physique Théorique des Liquides, Université Pierre et Marie Curie, 4, place Jussieu, 75252 Paris Cedex, 05 France

The parking-lot model provides a qualitative description of the main features of the phenomenology of granular compaction. We derive here approximate kinetic equations for this model, equations that are based on a 2−parameter generalization of the statistical-mechanical formalism first proposed by Edwards and coworkers. We show that history-dependent effects, such as memory and Kovacs effects, are captured by this approach.

1. Introduction

The term of "glassy-dynamics" is now commonly used to describe out-of-equilibrium systems that display such generic features as very slow kinetics that prevent the system from reaching equilibrium in any reasonable experimental timescale, aging phenomena, history-dependent processes like hysteresis and memory effects. Among such systems are the "not-too-strongly" vibrated granular materials[1–5]. In the recent years, there has been a surge of research activity in this field[6–8], partly driven by the goal of providing a statistical-mechanical description of these out-of-equilibrium situations[9–12][13–19].

In this note, we consider an approximate statistical-mechanical description of the parking-lot-model (PLM) for vibrated granular materials[2, 3, 20–24] that is based on the formalism proposed by Edwards and coworkers[9–12]. Despite its simplicity, the one-dimensional model of random adsorption-desorption of hard particles (PLM) has the merit of being a microscopic, off-lattice model that mimics many features of the compaction of a vibrated column of grains. It also has, we hope, a didactic value as to the nature of several canonical characteristics of "glassy dynamics". Many of the properties of the model, that can be obtained either analytically or by computer simulation, have been already described in the literature[2, 3, 20–26]. Our main focus here is on memory effects, including the so-called Kovacs effect first observed in glassy polymers[27, 28](see also[29–32]), and on the ingredients that are needed in a statistical-mechanical description to account for such effects.

2. The model and its properties

The parking-lot model is one-dimensional process in which hard rods of length σ are deposited at random positions on a line at rate k_+ and are inserted successfully only if they do not overlap with previously adsorbed particles; otherwise they are rejected.

Moreover, all deposited particles can desorb, i.e., be removed from the line at random with a rate k_-. For convenience, the unit time is set to $1/k_+$, and the unit length to σ. With this choice of units, the only control parameter in the model is $K = k_+/k_-$. When desorption is forbidden ($k_- = 0$), the model corresponds to the purely irreversible one-dimensional random sequential adsorption (RSA) process[33, 34], also known as the car parking problem, and all the properties of the system as a function of time can be obtained exactly. Connection to the compaction of a vibrated column of grains is made by considering the line as average layer (a 2-dimensional model would be more realistic, but the qualitative behavior would not be altered), the time as the number of taps, and $1/K$ as the tapping strength that controls the fraction of particles ejected from the layer at each tap.

When $1/K$ is not strictly equal to zero, adsorption and desorption are competing mechanisms that drive the system to a steady state corresponding to an equilibrium fluid of hard rods at a constant activity $1/K$. All the properties of the steady-state can also be obtained exactly.

The compaction kinetics of the parking-lot model at constant K is described by

$$\left.\frac{\partial \rho}{\partial t}\right|_K = \Phi(t) - \frac{\rho(t)}{K} \tag{1}$$

where $\rho(t)$ the density of hard rods on the line at time t and $\Phi(t)$ is the fraction of the line that is available at time t for inserting a new particle, i.e., the probability associated with finding an interval free of particles (a "gap") of length at least 1. The quantities ρ and Φ can be obtained from the 1−gap distribution function $G(h,t)$ which is the density of gaps of length h at time t via a number of "sum rules":

$$\rho(t) = \int_0^\infty dhG(h,t) = 1 - \int_0^\infty dhhG(h,t), \tag{2}$$

$$\Phi(t) = \int_1^\infty dh(h-1)G(h,t). \tag{3}$$

The 1−gap distribution function $G(h,t)$ obeys a kinetic equation that also involves 2−gap distribution function. Similarly, the kinetic equation of the 2-gap distribution involves the 3−gap distribution function, and so on[25]. The resulting infinite hierarchy of coupled equations cannot in general be solved analytically (exceptions are the RSA, when $k_- = 0$ and equilibrium when $t \to +\infty$).

The description of the kinetics of the parking lot model at large, constant K (i.e. small, constant tapping strength) can be summarized as follows: the density increases monotonically during the process, and the kinetics can be considered as a succession of four different regimes (see Fig. 1): during a first stage, the density increases rapidly until a value of around 0.65, and it is followed by an algebraic approach toward the saturation density, $\rho_{JL} = 0.747\ldots$ of the model without desorption (RSA); around this density there is a crossover to a still slower $1/\ln(t)$ compaction regime that is reminiscent of what is exactly observed when $1/K \to 0$, and finally there is an exponential approach to the steady-state (equilibrium) density with a rate $\Gamma \sim \frac{(\ln(K))^3}{K^2}$. For the two first regimes, desorption has a negligible effect and only the two last regimes are relevant

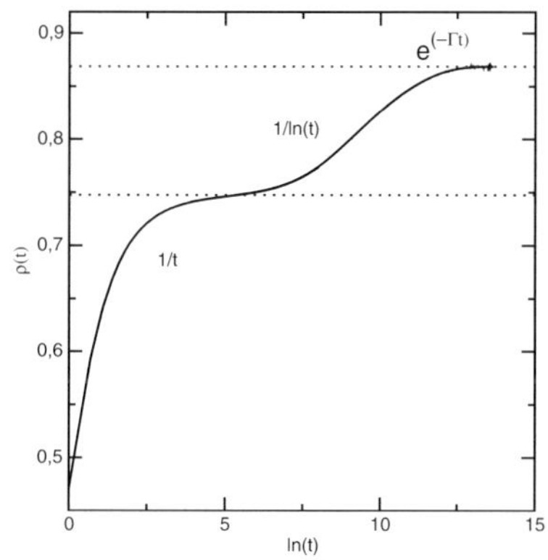

Figure 1. Logarithmic-linear plot of the density versus time for the parking-lot model at a constant tapping strength $1/K = 0.0002$

for the compaction kinetics of vibrated granular materials (although the approach to equilibrium may be prohibited when K is very large).

Experimentally, it has been observed that granular compaction exhibits history-dependent phenomena. In particular, Josserand *et. al*[4] have shown that memory effects are seen when changing the tapping strength during the compaction kinetics. Similar effects have been found in the parking-lot model, when the tapping strength $1/K$ is switched at a given time to a larger (or smaller) value: see Fig. 2

The obvious lesson that one can draw such memory effects is that whereas equilibrium is fully described by one thermodynamic parameter, the density ρ, out-of-equilibrium situations require at least one additional "thermodynamic" parameter. This parameter cannot simply be the tapping strength since the system can be found in states characterized by the same tapping strength $1/K$ and the same density ρ, that nonetheless evolve in different ways under further tapping with the same strength $1/K$: this is illustrated in Fig. 2b. A natural candidate for an additional thermodynamic parameter is the available line fraction Φ. One can indeed check that the two states described above do correspond to distinct values of Φ.

3. Statistical-mechanical formalism with two thermodynamic parameters

Following the ideas put forward by Edwards and coworkers[9–11], we consider a statistical-mechanical description of the system in which all possible microstates characterized by a (small) number of fixed macroscopic quantities are assumed to be equiprobable ("flat" or microcanonical distribution). In addition to fixing the density ρ, the parameter originally

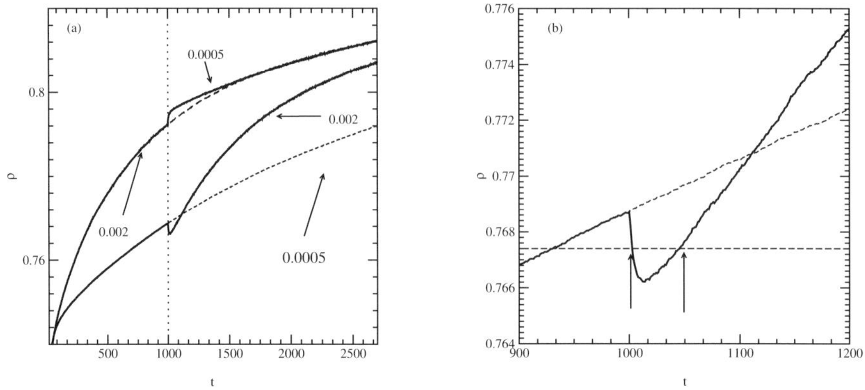

Figure 2. Memory effect: (a) density versus time when the tapping stength is switched from $5.e^{-4}$ to $2.e^{-3}$ (lower full curve) at $t = 1000$ and from $2.e^{-3}$ to $5.e^{-4}$ (upper full curve). The dashed curves correspond to a constant strength, $5.e^{-4}$ (lower curve) and $2.e^{-3}$ (upper curve); (b) Zoom up on the region around $t = 1000$ when one switches from $1/K = 5.e^{-4}$ to $2.e^{-3}$. The arrows denote two points at the same density and same tapping strength $1/K$.

selected in the compactivity-based description of granular media by Edwards and coworkers, we also constrain the microcanonical ensemble by fixing the available line fraction Φ to account for the above discussion.

Denoting by A the total length available for insertion of a particle center ($A = \Phi L$), the configurational integral with the constraints of fixed A, fixed system size L, and fixed number of particles N is obtained as[35]

$$Z(L,N,A) = \int_0^L \cdots \int_0^L \prod_{i=1}^{N} dh_i \delta\left(L - N - \sum_{i=1}^{N} h_i\right) \delta\left(A - \sum_{i=1}^{N} \theta(h_i - 1)(h_i - 1)\right), \quad (4)$$

which can be rewritten as

$$Z(L,N,A) = \int_C dz \int_{C'} dy \exp\left(L\left[z(1-\rho) + y\Phi + \rho \ln\left(\frac{z + y(1-\exp(-z))}{z(z+y)}\right)\right]\right) \quad (5)$$

where C and C' denote two closed contours. In the macroscopic limit, $N \to \infty$, $L \to \infty$, $A \to \infty$ with ρ and Φ fixed, one can use a saddle-point method to evaluate the integrals, which leads to

$$Z(L,N,A) \simeq \exp(Ls(\rho,\Phi)) \quad (6)$$

where the entropy density $s(\rho,\Phi)$ is given by

$$s(\rho,\Phi) = (1-\rho)z + y\Phi + \rho\ln\left(\frac{z+y(1-\exp(-z))}{z(z+y)}\right), \tag{7}$$

with $z \equiv z(\rho,\Phi)$ and $y \equiv y(\rho,\Phi)$ solutions of the two coupled equations

$$\left(\frac{1-\rho}{\rho}\right) = \frac{1}{z} + \frac{1}{z+y} - \frac{1+ye^{-z}}{z+y(1-e^{-z})}, \tag{8}$$

$$\frac{\Phi}{\rho} = \frac{1}{z+y} - \frac{1-e^{-z}}{z+y(1-e^{-z})}. \tag{9}$$

The gap distribution functions can also derived along the same lines[35], which leads to

$$G_{Ed}(h;\rho) = \begin{cases} \rho\dfrac{z(z+y)}{z+y\,(1-e^{-z})}e^{-zh} & \text{for } h<1, \\ \rho\dfrac{z(z+y)}{z+y\,(1-e^{-z})}e^{-(zh+y(h-1))} & \text{for } h>1. \end{cases} \tag{10}$$

It can also be shown that the multi-gap distribution functions satisfy a factorization property, e.g., $G_{Ed}(h,h';\rho,\Phi) = G_{Ed}(h;\rho,\Phi)G_{Ed}(h';\rho,\Phi)$. Note that the 1-gap distribution function is a piecewise continuous function that obeys the exact sum rules, Eqs. (2) and (3).

A detailed comparison between the predictions of this statistical-mechanical treatment and simulation data *at the same ρ and Φ* can be found in Ref.[35]. The conclusion is that although not exact, and even missing some qualitative features in the limiting case of a purely irreversible RSA process, the approach provides an overall good description of the data. (It becomes of course exact in the steady state since this latter corresponds to an equilibrium situation.)

4. Approximate description of the compaction kinetics

A quasi-thermodynamic approach is useful because it allows to predict the structure of the system (correlation functions) as well as the fluctuations at any given state point (here assumed to characterized by ρ and Φ). In the cases where phase transitions occur, it can also provide interesting constraints and relations between the parameters characterizing the phases at the transition or help to determine the limit of stability of the phases in mean-field like treatments. However, one still faces the problem of predicting the state of the system *for a given protocol and a given history*: in the simplest case, for a given time t and a given tapping strength $1/K$.

We thus want to push the Edwards formalism one step further. Predicting the trajectory made by the system in the (ρ,Φ) diagram for a given protocol amounts to determine a system of equations relating ρ and Φ with t and K. Eq. (1) is one such equation. To obtain another equation for the evolution of Φ, we start for the exact kinetic equation for the 1−gap distribution function for $h \geq 1$:

$$\left.\frac{\partial G(h,t)}{\partial t}\right|_K = -(h-1)G(h,t) + 2\int_{h+1}^{\infty} dh' G(h',t) - \frac{2G(h,t)}{K} + \frac{1}{K}\int_0^{h-1} dh' G(h',h-1-h',t),$$

(11)

where $G(h,h',t)$ is the 2-gap distribution function that satisfies the "sum rule"

$$\int_0^{\infty} dh' G(h,h',t) = \rho(t)G(h,t). \tag{12}$$

Note that the approximate 2−gap distribution function $G_{Ed}(h,h')$ satisfies the above sum rule. By multiplying Eq. (11) by $(h-1)$, integrating over h from 1 to ∞, and using the sum rules, Eqs. (2), (3) and (12), one obtains an exact equation for the evolution of Φ:

$$\left.\frac{\partial \Phi(t)}{\partial t}\right|_K = \frac{2}{K}(1-\rho(t)-\Phi(t)) - \int_1^{\infty} dh(h-1)^2 G(h,t) + \int_2^{\infty} dh(h-2)^2 G(h,t). \tag{13}$$

If one now inserts the approximate expression of the 1-gap distribution function, Eq. (10), in the above equation, one gets

$$\left.\frac{\partial \Phi}{\partial t}\right|_K = \frac{2(1-\rho(t)-\Phi(t))}{K} - 2\Phi(t)\frac{1-e^{-(y(t)+z(t))}}{y(t)+z(t)}. \tag{14}$$

Eqs. (1), (8), (9) and (14) form a closed set of equations whose solution for given initial conditions completely characterizes the system and its evolution.

We first test the accuracy of the above approximate kinetic description in two limiting cases for which the exact solution is known: the purely irreversible RSA case ($1/K = 0$) and the approach to the steady state ($t \to +\infty$) for a given rate K. The steady state corresponds to the solution $z_\infty = \frac{\rho_\infty}{1-\rho_\infty}$, $y_\infty = 0$, where ρ_∞ is the equilibrium density of hard rods at constant activity $1/K$: $\rho_\infty = K\Phi_\infty$ with $\Phi_\infty = (1-\rho_\infty)\exp\left(\frac{-\rho_\infty}{1-\rho_\infty}\right)$; it is thus the exact result. The approach to the steady state is obtained by linearizing the kinetic equations around the equilibrium solution. One then obtains the following coupled linear differential equations :

$$\left.\frac{\partial y(t)}{\partial t}\right|_K = -2\frac{(z_\infty+1)(e^{-2z_\infty}-1)+2(1+z_\infty^2)e^{-2z_\infty}}{z_\infty(2e^{z_\infty}-z_\infty^2-2z_\infty-2)} y(t) \tag{15}$$

$$\left.\frac{dz(t)}{dt}\right|_K = -\frac{6e^{z_\infty}-2z_\infty^2-6-8z_\infty+(2+z_\infty^2)z_\infty e^{-z_\infty}}{z_\infty(2e^{z}_\infty-z_\infty^2-2z_\infty-2)} y(t) - \frac{z_\infty+1}{z_\infty}(z(t)-z_\infty). \tag{16}$$

The associated eigenvalues are all negative and that corresponding to the inverse of relaxation time goes for large K as

$$\tau^{-1} = \Gamma \sim \frac{\ln(K)^2}{3K}. \tag{17}$$

The relaxation time is of the same order of magnitude as that obtained in the simple adiabatic approximation[3] and is much smaller than the exact result (see above). The error comes from the inability of the Edwards approximation to account for the non-exponential behavior of the 1−gap distribution function for $0 \leq h \leq 1$.

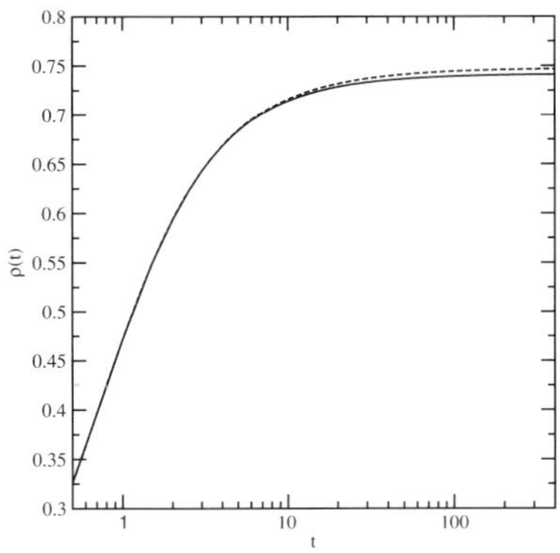

Figure 3. Density versus time for the model without desorption (RSA): the full curve corresponds to the approximation, the dashed curve is the exact result.

One the other hand the approximate kinetic description is very good in the purely irreversible case, $1/K = 0$, (but it still misses some features, as discussed above and in Ref.[35]). Then Eqs. (1) and (14) simplify to

$$\frac{\partial \rho}{\partial t} = \Phi(t) \tag{18}$$

$$\frac{\partial \Phi}{\partial t} = -2\Phi(t)\frac{1 - e^{-(y(t)+z(t))}}{y(t) + z(t)}, \tag{19}$$

where y and z are related to ρ and Φ by Eqs. (8) and (9). It is easy to show that at long times $z(t)$ goes to a finite limit whereas $y(t)$ goes to ∞ as t. As a result the kinetics approaches a non-trivial jamming limit with an algebraic $1/t$ behavior. The numerical solution of the approximate equation shows that the saturation density at the jamming limit $\rho_{JL}^{Ed} \simeq 0.7422$ is very close to the exact value, $\rho_{JL} = 0.74759\ldots$ The overall agreement with the exact density evolution is very good, as shown in Fig. 3.

5. Memory and Kovacs Effects

Finally, we apply the approximate kinetic equations to the description of memory effects. By introducing Φ as the second state variable, one expects to obtain a response to a sudden change of the tapping strength that captures these memory effects. This is indeed illustrated in Fig. 4a that shows how the density evolves when K is changed at $t = 60$ from $K = 500$ to $K = 5000$ and from $K = 500$ to $K = 200$. At longer times, note that the density curve corresponding to the change from $K = 500$ to $K = 200$ crosses two times

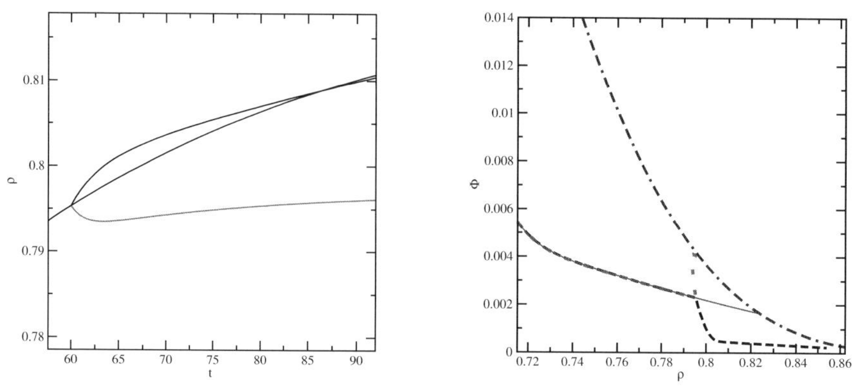

Figure 4. (a) Approximate time evolution of the density for three cases: $K = 500$ (full curve), $K = 500$ from $t = 0$ to $t = 60$ and $K = 200$ for $t > 60$ (dotted curve) and $K = 500$ from $t = 0$ to $t = 60$ and $K = 5000$ for $t > 60$ (dashed curve). (b)Parametric plot of insertion probability Φ versus density for the three above cases. For completeness, the equilibrium insertion probability is added (dash-dot curve).

the density curve calculated when $K = 200$ is kept constant along the process. This is the signature of a Kovacs-like effect which is more clearly displayed in the parametric plot of Fig. 4b: instead of going monotonically to the final value, the density decreases first and passes by a minimum after increasing again. This non-monotonic behavior of the density (or the volume) is precisely the Kovacs effect. The same phenomenon is illustrated for different values of the tapping strength in Figs 5a and b: K is switched from 5000 to 500 or from 5000 to 2000 after a time $t = 250$. In the former case, the density reached at $t = 250$ when K is changed is higher than the final (equilibrium) density reached at $K = 500$; nonetheless, the density starts by first decreasing to a value less than the final density until it reaches a minimum and finally increases again. On the other hand, no minimum is observed in the latter case.

The Kovacs effect is traditionally represented as a "hump" in the volume as a function of time[27, 30, 31]. The position and the height of the hump vary with the waiting time, i.e., the time spent at the initial tapping strength (recall that the process of compaction is always out of equilibrium), and with the amplitude of the shift in the tapping strength (or in other glassy systems, the shift in temperature). The same behavior is found here. It is illustrated in Fig. 6 where we show the evolution of the inverse density for different protocols to reach the equilibrium steady state at $K = 500$: in the upper curve, K is switched from 5000 to 500 at $t_w = 240$, in the intermediate curve K is switched from 2000 to 500 at $t_w = 169$ and in the lower curve $K = 1000$ at $t_w = 139$. The waiting time t_w is

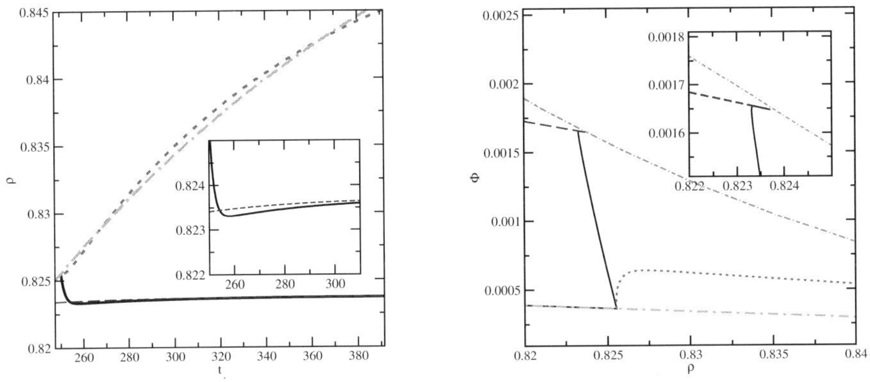

Figure 5. (a)Approximate time evolution of the density for three cases:$K = 500$ (lower curve), $K = 5000$ (dot-dashed curve), $K = 5000$ from $t = 0$ to $t = 250$ and $K = 500$ for $t > 250$ (full curve) and $K = 5000$ from $t = 0$ to $t = 250$ and $K = 2000$ for $t > 60$ (dotted curve). (b)Parametric plot of the insertion probability Φ versus density for the four above cases: For completeness,the equilibrium insertion probability is added (dashed curve).

chosen that the density reached at that time is equal to final density, $\rho_{eq}(K = 500)$. One observes that the height of the hump increases as the amplitude of the shift increases. The Kovacs effect, also observed in the simulation data (not shown here), results from competing trends that are qualitatively well accounted for by the present description in terms of two thermodynamic parameters. Including Φ in the statistical-mechanical treatment is crucial for reproducing such behavior.

6. Conclusion

We have applied a two-parameter statistical-mechanical formalism inspired by the work of Edwards and coworkers to describe the compaction kinetics of the parking-lot model. The approximation gives a fair description of the kinetics, although it underestimates the relaxation time characteristic of the final approach to equilibrium. Inclusion of a second thermodynamic parameter allows one to qualitatively reproduce experimentally observed memory effects in granular compaction and to predict a Kovacs effect similar to what is observed in many glassy systems.

REFERENCES

[1] J. B. Knight, C. G. Fandrich, C. N. Lau, H. M. Jaeger, and S. R. Nagel, Phys. Rev. E **51**, 3957 (1995).

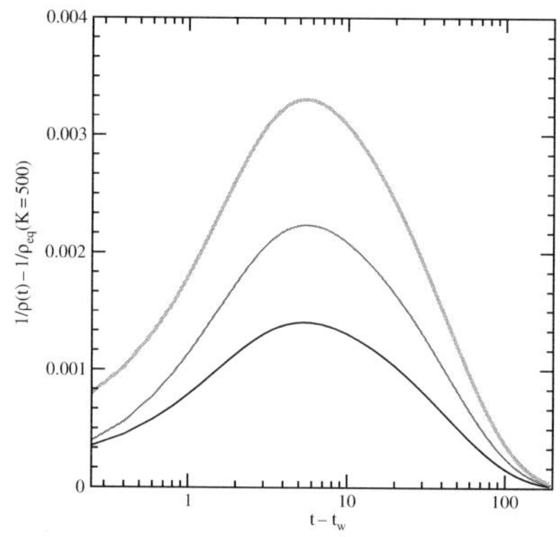

Figure 6. Kovacs effect in the approximate description of the parking-lot model : $1/\rho(t) - 1/\rho_{eq}(K = 500)$ versus $t - t_w$. In the upper curve K is switched from $K = 5000$ to $K = 500$ at $t_w = 240$, in the intermediate curve K is switched from $K = 2000$ to $K = 500$ at $t_w = 169$, and in the lower curve, K is switched from 1000 to 500 at $t_w = 139$.

[2] E. R. Nowak, J. B. Knight, E. Ben-Naim, H. M. Jaeger, and S. R. Nagel, Phys. Rev. E **57**, 1971 (1998).

[3] E. Ben-Naim, J. B. Knight, E. R. Nowak, H. M. Jaeger, and S. R. Nagel, Physica D **123**, 380 (1998).

[4] C. Josserand, A. Tkachenko, D. Mueth, and H. Jaeger, Phys. Rev. Lett. **85**, 3632 (2000).

[5] P. Philippe and D. Bideau, Europhys. Lett. **60**, 677 (2002).

[6] J.-P. Bouchaud, in *Slow Relaxations and Nonequilibrium Dynamics in Condensed Matter*, Vol. 77 of *Les Houches-Ecole d'Ete de Physique Theorique*, edited by J. L. Barrat, M. Feigelman, J. Kurchan, and J. Dalibard (Springer-Verlag, Berlin, 2003).

[7] J.-P. Bouchaud, L. Cugliandolo, J. Kurchan, and M. Mezard, in *Spin Glasses and Random Fields* (World Scientific, Singapore, 1998), p. 161.

[8] A. Liu and S. Nagel, *Jamming and Rheology: constrained dynamics* (Taylor and Francis, London, 2001).

[9] A. Mehta and S. Edwards, Physica A **157**, 1091 (1989).

[10] S. Edwards and R. Oakeshott, Physica A **157**, 1080 (1989).

[11] S. Edwards, in *Granular Matter: An Interdisciplinary Approach*, edited by A. Mehta (Springer-Verlag, New York, 1994).

[12] S. Edwards and D. Grinev, Phys. Rev. E **58**, 4758 (1999).

[13] A. Barrat, J. Kurchan, V. Loreto, and M. Sellito, Phys. Rev. Lett. **85**, 503 (2000).

[14] A. Barrat, V. Colizza, and V. Loreto, Phys. Rev. Lett. **66**, 011310 (2002).

[15] D. Dean and A. Lefèvre, Phys. Rev. Lett. **90**, 198301 (2003).
[16] A. Lefèvre, J. Phys. A.: Math. Gen. **35**, 9037 (2002).
[17] J. Berg, S. Franz, and M. Sellitto, Eur. Phys. J. B **26**, 349 (2002).
[18] A. Coniglio, A. Fierro, and M. Nicodemi, cond-mat/0305175 (2003).
[19] A. Prados and J. Brey, Phys. Rev. E **66**, 041308 (2002).
[20] X. Jin, G. Tarjus, and J. Talbot, J. Phys. A.: Math. Gen. **27**, L195 (1994).
[21] P. L. Krapivsky and E. Ben-Naim, J. Chem. Phys. **100**, 6778 (1994).
[22] A. Kolan, E. Nowak, and A. Tkachenko, Phys. Rev. E **59**, 3094 (1999).
[23] J. Talbot, G. Tarjus, and P. Viot, J. Phys. A.: Math. Gen. **32**, 2997 (1999).
[24] J. Talbot, G. Tarjus, and P. Viot, Eur. Phys. J. E **5**, 445 (2001).
[25] J. Talbot, G. Tarjus, and P. Viot, Phys. Rev. E **61**, 5429 (2000).
[26] J. Talbot, G. Tarjus, and P. Viot, Fractals **11**, 185 (2003).
[27] A. Kovacs, Fortschr. Hochpolym. Forsch. **3**, 394 (1963).
[28] L. Struik, *Physical Aging in Amorphous Polymers and Other Materials* (Elsevier, New York, 1978).
[29] L. Berthier and J.-P. Bouchaud, Phys. Rev. B **66**, 054404 (2002).
[30] S. Mossa and F. Sciortino, cond-mat/0305526 (2003).
[31] E. Bertin, J.-P. Bouchaud, J.-M. Drouffe, and C. Godrèche, cond-mat/0306089 (2003).
[32] A. Buhot, cond-mat/0310311 (2003).
[33] J. Evans, Rev. Mod. Phys. **65**, 1281 (1993).
[34] J. Talbot, G. Tarjus, P. V. Tassel, and P. Viot, Colloids and Surf. A: **165**, 287 (2000).
[35] G. Tarjus and P. Viot, cond-mat/0307267 (2003).

Unifying Concepts in Granular Media and Glasses
A. Coniglio, A. Fierro, H.J. Herrmann and M. Nicodemi (editors)

Statistical Mechanics of jamming and segregation in granular media

M. Nicodemi[a], A. Coniglio[a], A. de Candia[a], A. Fierro[a], M. Pica Ciamarra[a], and M. Tarzia[a]

[a]Dipartimento di Fisica, Università di Napoli "Federico II", INFM, Unità di Napoli, Complesso Universitario Monte Sant'Angelo, Via Cinthia, I-80126, Napoli, Italy

In the framework of schematic hard spheres lattice models we discuss Edwards' Statistical Mechanics approach to granular media. As this approach appears to hold here to a very good approximation, by analytical calculations of Edwards' partition function at a mean field level we derive the system phase diagram and show that "jamming" corresponds to a phase transition from a "fluid" to a "glassy" phase, observed when crystallization is avoided. The nature of such a "glassy" phase turns out to be the same found in mean field models for glass formers. In the same context, we also briefly discuss mixing/segregation phenomena of binary mixtures: the presence of fluid-crystal phase transitions drives segregation as a form of phase separation and, within a given phase, gravity can also induce a kind of "vertical" segregation, usually not associated to phase transitions.

1. Introduction

We review some recent results in the study of glasses and dense granular materials obtained with the aid of schematic lattice models, simple enough to allow a full numerical and analytical investigation. Glasses and granular media exhibit deep similarities in their "jamming" behaviors, but although the idea of a unified description is emerging (see Silbert, O'Hern, Liu and Nagel contribution to this volume), the precise nature of jamming in non-thermal systems, such as granular media, and the origin of its close connections to glassy phenomena in thermal ones are still open and very important issues.

An important conceptual open problem concerning granular media, is the absence of an established theoretical framework where they might be described. Edwards [1] (see Edwards, Brujić, and Makse contribution to this volume) proposed a solution to such a problem by introducing the hypothesis that time averages of a system, exploring its mechanically stable states subject to some external drive (e.g., "tapping"), coincide with suitable ensemble averages over its "jammed states". We discuss here some recent results validating and generalizing Edwards' proposal (see also Dean and Lefevre, and Tarjus and Viot contributions). In particular, within this approach we show that mean field models for granular media undergo a phase transition from a (supercooled) "fluid" phase to a "glassy" phase, when their crystallization transition is avoided. The nature of such a "glassy" phase results to be the same found in mean field models for glass formers: a discontinuous one step Replica Symmetry Breaking phase preceded by a dynamical freezing point. These results are supported by Monte Carlo (MC) "tap dynamics" simulations which, in the region of low MC shaking amplitudes, show a pronounced jamming similar to the one found in experiments on granular media (see Bideau, Philippe, Ribière and Richard, and Caballero, Lindner, Ovarlez, Reydellet, Lanuza and Clement, and D'Anna and Mayor contributions to the volume). As an application of Edwards' approach to powders we also briefly discuss segregation/mixing phenomena in these systems and discuss close correspondences with experiments (see Reis, Mullin and Ehrhardt contribution).

Sec.2 gives an short introduction to the phenomenology of granular materials relevant to the present purpose. Sec.3 briefly reviews the essential lines in Edwards' Statistical Mechanics of powders and discuss some correspondence with other systems "frozen" at $T = 0$, such as glasses in their inherent structures. In Sections 4 to 6 we describe our results obtained in the study of hard spheres lattice models of granular materials, in order to check and apply Edwards' Statistical Mechanics to jamming and segregation. These are obtained by computer simulations and by analytic calculations with the generalized Bethe-Peierls method, or "cavity method", recently developed by Mézard and Parisi [2]. Finally, we draw some conclusions.

2. Granular Media

Granular media are large conglomerations of discrete macroscopic particles. Ubiquitous in the world around us, in many respect they behave differently from any of the other familiar forms of matter - solids, liquids, or gases (for a review see [3]). The static and dynamic properties of granular media, which are strongly dissipative systems, are not affected by temperature, because thermal fluctuations are usually negligible. In sand the potential energy of a grain raised a distance equal to its own diameter can be 10^{12} times $k_B T$ at room temperature; therefore the temperature of the external bath can be considered zero and these media called *non-thermal*. As the system cannot explore its phase space, unless perturbed by external forces (such as shaking or tapping), it is frozen, at rest, in its mechanically stable microstates.

Even though granular media may form crystalline packings, in most cases they are found at rest in disordered configurations, characterized by "fluid" like distribution functions. Actually, gently shaken granular media exhibit a strong form of "jamming" [4–6], i.e., an exceedingly slow dynamics, which shows deep connections to "freezing" phenomena observed in many thermal systems such as glass formers [7,8]. It was discovered by the Chicago group [4] that under tapping the density of granular media tends to increase very slowly as a function of the number of shakes, in a stretched exponential [6] or even logarithmic [4] way. These systems have typical "aging phenomena" along with huge relaxation times diverging á la Arrhenius or Vogel and Fulcher [4–6], similarly to glass formers in the freezing region.

Here, we discuss the nature of jamming in non-thermal systems [8,9], and the origin of its close connections to glassy phenomena in thermal ones, in the framework of the Statistical Mechanics of powders introduced by Edwards [1,10]: we derive the phase diagram of granular media and explain in a quantitative way their similarities and differences with glass formers. The introduction of a Statistical Mechanics for powders is grounded on observations from experiments [4,6] and simulations [14,10] that, as much as in thermal systems, their macroscopic properties at stationarity are characterized by a few control parameters and their macrostates correspond to a huge number of microstates.

As an application of the Statistical Mechanics of powders, we also consider the intriguing phenomenon of segregation: in presence of shaking a granular system is not randomized, but its components tend to separate [15]. An example is the so called "Brazil nut" effect (BNE) where, under shaking, large particles rise to the top and small particles move to the bottom of the container. Interestingly, by changing grains sizes or mass ratio or shaking amplitudes a crossover towards a "reverse Brazil nut" effect (RBNE) was more recently discovered [20] where small particles segregates to the top and large particles to the bottom. Several mechanisms have been proposed to explain these phenomena which, although of deep practical and conceptual relevance, are still largely unknown[15]. Geometric effects, such as "percolation" [16] or "reorganization" [17,18], are known to be at work since, in a nutshell, small grains appear to filter beneath large ones. "Dynamical" effects, such as convection [19] or inertia [21], were shown to play a role as well. Recent simulations and experiments have, however, outlined that segregation phenomena can involve "global" mechanisms, such as "condensation" [20] or, more generally, "phase separation" [22]. We focus on these properties here.

3. Approaches á la Edwards to Statistical Mechanics of granular media

In the Statistical Mechanics of powders introduced by Edwards [1] it is postulated that the system at rest (i.e., not in the "fluidized" regime) can be described by suitable ensemble averages over its "mechanically stable" states (called here "inherent states"). Edwards proposed a method to individuate the probability, P_r, to find the system in its inherent state r, under the assumption that these mechanically stable states have the same a priori probability to occur. The knowledge of P_r has the conceptual advantage to substitute *time* with *ensemble averages*, allowing the description of the system properties by use of few basic theoretical concepts, as in thermodynamics. A possible approach to find P_r is as follows [1,10]. P_r is obtained as the maximum of the entropy,

$$S = -\sum_r P_r \ln P_r \tag{1}$$

with the macroscopic constraint, in the case of the canonical ensemble, that the system energy, $E = \sum_r P_r E_r$, is given. This assumption leads to the Gibbs result:

$$P_r \propto e^{-\beta_{conf} E_r} \tag{2}$$

where β_{conf} is a *Lagrange multiplier*, called *inverse configurational temperature*, enforcing the above constraint on the energy:

$$\beta_{conf} = \frac{\partial S_{conf}}{\partial E} \qquad S_{conf} = \ln \Omega_{IS}(E) \tag{3}$$

Here, $\Omega_{IS}(E)$ is the number of inherent states with energy E. Thus, summarizing, the system at rest has $T_{bath} = 0$ and $T_{conf} = \beta_{conf}^{-1} \neq 0$.

These basic considerations, to be validated by experiments or simulations, settle a theoretical Statistical Mechanics framework to describe granular media. Consider, for definiteness, a system of monodisperse hard spheres of mass m. In the system whole configuration space Ω_{Tot}, we can write Edwards' generalized partition function as:

$$Z = \sum_{r \in \Omega_{Tot}} \exp(-\mathcal{H}_{HC} - \beta_{conf} mgH) \cdot \Pi_r \tag{4}$$

where $\mathcal{H}_{HC}$ is the hard core interaction between grains, mgH is the gravity contribution to the energy (H is particles height), and the factor Π_r is a projector on the space of "mechanically stable" states (i.e., inherent states space) Ω_{IS}: if $r \in \Omega_{IS}$ then $\Pi_r = 1$ else $\Pi_r = 0$. Usual Statistical Mechanics, where β_{conf}^{-1} is identified with T_{bath}, is recovered in the case where $\Pi_r = 1\ \forall r$.

In order to test Edwards proposal one introduces a dynamics (such as "trapping") that allows the system to explore the inherent states space. Using this dynamics one has first to check that at stationarity the system properties do not depend on the details of the dynamical history, i.e., a "thermodynamic" description is indeed possible. Then one must check that time averages obtained using such a dynamics compare well with ensemble averages over the distribution Eq.(2).

3.1. Configurational entropy of granular media and glasses frozen at $T = 0$

Edwards approach could also be relevant for a supercooled liquid quenched at very low temperature (about zero). In this case the system remains blocked in its "inherent structures", i.e., the local minima of the potential energy in the particle configuration space [23,24,11,12] (by analogy with the glass terminology we call "inherent states" the mechanically stable states in granular materials).

In glasses, in the inherent structure approach [23,24,11,12], the configurational entropy (and consequently the configurational temperature) associated to the number of inherent structures corresponding to a given energy, E, can be defined too. In this context two different ways were essentially used to allow the system explore its inherent structure space. One way is by quenching the system over and over from an equilibrium temperature, T, to zero temperature [23,11,12]. Using this procedure, Sciortino *et al.* [11] found that the configurational temperature numerically coincides with the equilibrium temperature T, provided that T is low enough. Another way to visit the inherent structures is by letting the system aging in contact with an almost zero bath temperature, T_{bath}. During the aging process an effective temperature, T_{dyn}, can be defined via the off-equilibrium extension of the Fluctuation-Dissipation ratio [26]. It happens that in mean field models [25,26] this effective temperature coincides with the configurational temperature, T_{conf}. This was also shown in Ref. [28], in the limit $T_{bath} \to 0$, for a class of finite dimensional systems. The connection between Edwards approach for granular media and the results in glass theory has been pointed out in [27,28,14,29,10].

A second procedure, called "tap dynamics", similar to that used in the compaction of real granular materials was instead used in [29,10,30–33,7]. In particular in Ref. [29,10] each tap consists in raising the bath temperature to a value T_Γ and, after a lapse of time τ_0, quenching it back to zero. By cyclically repeating the process the system explores the space of the inherent states. Once the stationary state is reached a temperature, T_{fd}, is defined via the equilibrium Fluctuation-Dissipation relation, and if Edwards' assumption applies, T_{fd} coincides with the configurational temperature. This was verified for different

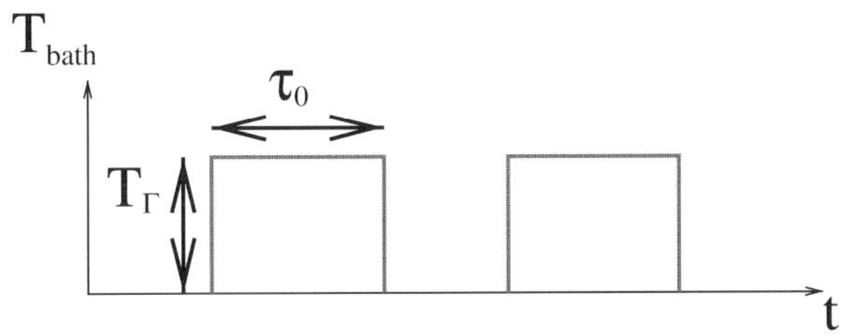

Figure 1. Our lattice models for granular media are subject to a Monte Carlo dynamics made of "taps" sequences. A "tap" is a period of time, of length τ_0 (the tap duration), during which the system evolves at a finite bath temperature T_Γ (the tap amplitude); after each "tap" the system evolves at $T_\Gamma = 0$ and reaches an inherent state.

finite dimensional models [29,10]. Many other studies discussing Edwards' approach in this perspective were also presented [30–32]. In the following sections we give a picture of results [29,10] about such a statistical mechanics approach in schematic models of granular media and glasses.

4. Lattice models for granular media

4.1. A monodisperse hard-sphere system under gravity

The simplest model for granular media we considered [10] is a system of hard-spheres of equal diameter $a_0 = 1$, subjected to gravity. We have studied this model on a lattice, constraining the centers of mass of the spheres on the sites of a cubic lattice (see inset in Fig. 3). The Hamiltonian of the system is:

$$\mathcal{H} = \mathcal{H}_{HC}(\{n_i\}) + gm \sum_i n_i z_i, \tag{5}$$

where the height of site i is z_i, $g = 1$ is gravity acceleration, $m = 1$ the grains mass, $n_i = 0, 1$ the usual occupancy variable (i.e., $n_i = 0$ or 1 if site i is empty of filled by a grain) and $\mathcal{H}_{HC}(\{n_i\})$ an hard-core interaction term that prevents the overlapping of nearest neighbor grains (this term can be written as $\mathcal{H}_{HC}(\{n_i\}) = J \sum_{\langle ij \rangle} n_i n_j$, where the limit $J \to \infty$ is taken).

The grains are subject to a dynamics made of a sequence of Monte Carlo "taps" (see Fig. 1): a single "tap" is a period of time, of length τ_0 (the tap duration), where particles can diffuse laterally, upwards with probability $p_{up} \in [0, 1/2]$, and downwards with probability $1 - p_{up}$. When the "tap" is off grains can only move downwards (i.e., $p_{up} = 0$) and the system evolves with $p_{up} = 0$ until it reaches a blocked configuration (i.e., an "inherent state") where no grain can move downwards without violating the hard core repulsion. The parameter p_{up} has an effect equivalent to keep the system in contact (for a time τ_0) with a bath temperature $T_\Gamma = mga_0/\ln[(1-p_{up})/p_{up}]$ (called the "tap amplitude"). The properties of the system are measured when this is in a blocked state. Time averages,

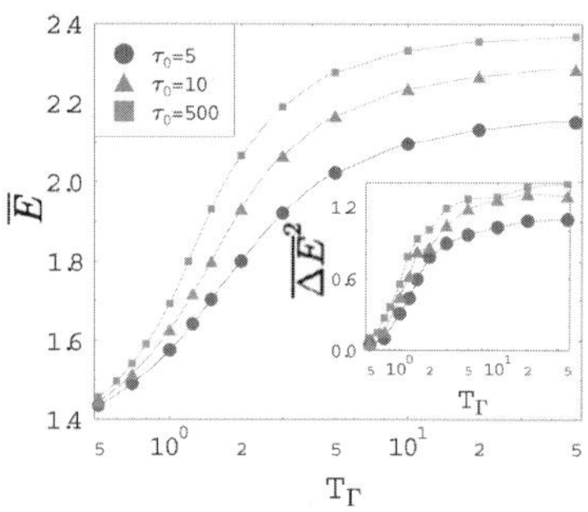

Figure 2. The time average of the energy, $\overline{E}$, and (**inset**) its fluctuations, $\overline{\Delta E}^2$, recorded at stationarity during a tap dynamics, as a function of the tap amplitude, T_Γ, in the 3D lattice monodisperse hard sphere model. Different curves correspond to sequences of tap with different values of the duration of each single tap, τ_0.

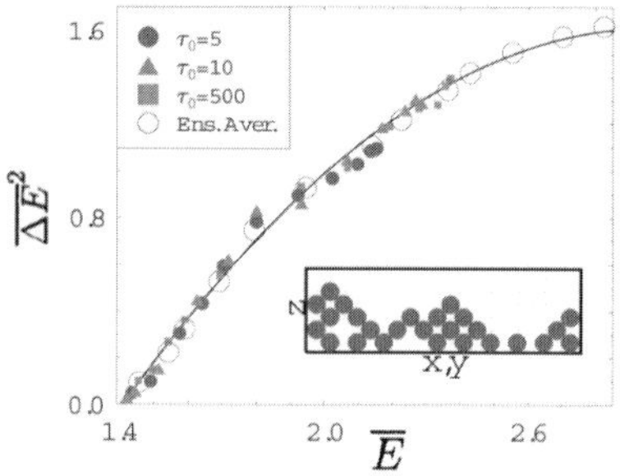

Figure 3. Time averages of energy fluctuations $\overline{\Delta E^2}$ plotted as function of the time average of energy $\overline{E}$. •, ▲ and ■ are time averages obtained with different tap dynamics. ○ are independently calculated ensemble averages according to Eq.(2). The collapse of the data obtained with different dynamics shows that the system stationary states are characterized by a *single* thermodynamic parameter. The agreement with the ensemble averages show the success of Edwards' approach to describe the system macroscopic properties.

therefore, are averages over the blocked configurations reached with this dynamics. Time t is measured as the number of taps applied to the system.

Under such a tap dynamics the systems reaches a stationary state where the Statistical Mechanics approach to granular media can be tested, and particularly Edwards hypothesis can be verified by comparing time averages to ensemble averages of Eq.(2).

4.2. The stationary states of the tap dynamics

During the tap dynamics, in the stationary state, the time average of the energy, $\overline{E}$, and its fluctuations, $\overline{\Delta E^2}$, are calculated. Figure 2 shows $\overline{E}$ (main frame) and $\overline{\Delta E}^2$ (inset) as function the tap amplitude, T_Γ, (for several values of the tap duration, τ_0). Since sequences of taps, with same T_Γ and different τ_0, give different values of $\overline{E}$ and $\overline{\Delta E}^2$, it is apparent that T_Γ is not the right thermodynamic parameter. On the other hand, if the stationary states are indeed characterized by a *single* thermodynamic parameter the curves corresponding to different tap sequences (i.e. different T_Γ and τ_0) should collapse onto a single master function, when $\overline{\Delta E}^2$ is parametrically plotted as function of $\overline{E}$. This is the case in the present model, where the data collapse is in fact found and shown in Fig. 3. This is a prediction that could be easily checked in real granular materials.

A technique to derive from raw data the thermodynamic parameter β_{fd} conjugated to E (apart from an integration constant, β_0), is through the usual *equilibrium* Fluctuation-

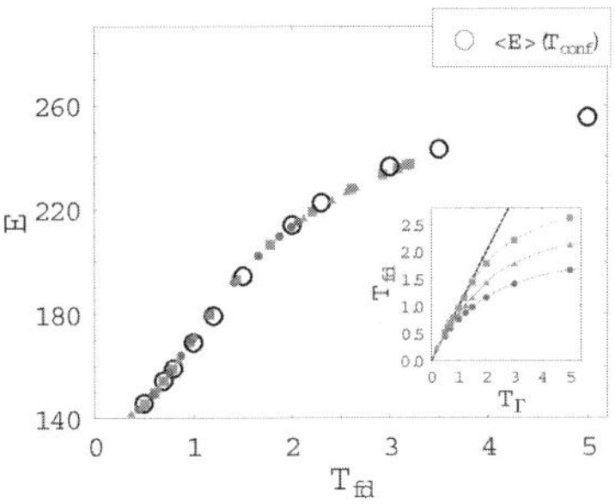

Figure 4. **Main frame** The time average $\overline{E}$ and the ensemble average over the distribution Eq.(2) $\langle E\rangle$, plotted respectively as a function of T_{fd} and T_{conf} (in units mga_0), in the 3D monodisperse hard-sphere system under gravity described in the text. Symbols are as in Fig. 3. Time averages over the tap dynamics and Edwards' ensemble averages coincide. **Lower Inset** The temperature $T_{fd} \equiv \beta_{fd}^{-1}$ defined by Eq.(6) as function of T_Γ (in units mga_0) for $\tau_0 = 500, 10, 5$ MCS (from top to bottom). The straight line is the function $T_{fd} = T_\Gamma$.

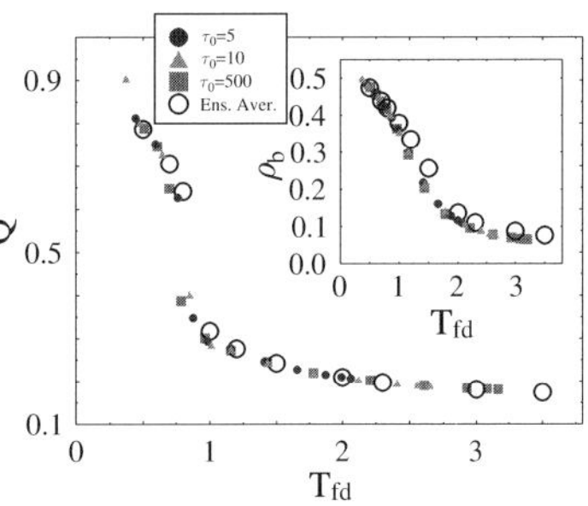

Figure 5. The density self-overlap function, Q, and (**upper inset**) the system density on the bottom layer, ρ_b, plotted as function of T_{fd} (in units mga_0), compared with the ensemble averages over the distribution Eq.(2) (the black empty circles), plotted as function of T_{conf} (in units mga_0), in the 3D monodisperse hard-sphere system under gravity. Also for this system there is a very good agreement between the independently calculated time averages over the tap dynamics and the statistical mechanics ensemble averages á la Edwards.

Dissipation relation:

$$-\frac{\partial \overline{E}}{\partial \beta_{fd}} = \overline{\Delta E^2}. \tag{6}$$

By integrating Eq.(6), $\beta_{fd} - \beta_0$ can be expressed as function of $\overline{E}$ or (for a fixed value of τ_0) as function of $\beta_\Gamma = 1/T_\Gamma$: $\beta_{fd} - \beta_0 \equiv g(\beta_\Gamma)$ (the constant β_0 can be determined as explained in [10]).

4.3. Edwards' averages

In Sect. 4.2 we have found that the fluctuations of the energy in the stationary state depend only on the energy, E, and not on the past history. More generally, we found [10] that all the macroscopic quantities we observed depend only on the energy, E, or on its conjugate thermodynamic parameter, β_{fd}, thus the stationary state can be genuinely considered a "thermodynamic state". In this case one can attempt to construct an equilibrium statistical mechanics, as originally suggested by Edwards [1].

To test Edward hypothesis one has to compare the time average of the energy, $\overline{E}(\beta_{fd})$, recorded during the taps sequences, with the ensemble average, $\langle E\rangle(\beta_{conf})$, over the distribution Eq.(2). Therefore we have independently calculated the ensemble average $\langle E\rangle$, as function of β_{conf}. Fig. 4 (see also Fig. 3) shows a very good agreement between $\langle E\rangle(\beta_{conf})$ and $\overline{E}(\beta_{fd})$ (notice that there are no adjustable parameters). In order to check the scenario in further details, we have also calculated the system density on the bottom layer, ρ_b, and the density self-overlap function Q: $Q = 1/N \sum_{i=1}^{N} n_i^{(1)} n_i^{(2)}$, where the sum is over all the lattice sites and $n^{(1)}$ and $n^{(2)}$ refer to two different copies of the systems, which are at stationarity under the same tap dynamics. We have verified that, when plotted as function of T_{fd}, both ρ_b and Q scale on a single master function (see Fig. 5). In Fig. 4 (inset) we also show the dependence of the configurational temperature T_{conf} on the parameters of the tap dynamics T_Γ and τ_0. Finally, we mention that we have also successfully tested Edwards scenario in an other model, the "frustrated lattice gas" [34,10], a system in the category of spin glasses.

4.4. Relaxation during MC "tap" dynamics

The MC tap dynamics, in both the "frustrated lattice gas" and hard sphere models, exhibits a rich structure in agreement with experimental findings [4,6]. In the region of small tap amplitudes, the system gets "jammed" and "memory" phenomena are observed, along with "irreversibility" effects [4,7]. In particular, it is interesting to consider correlation functions such as $C(t, t_w) = B(t, t_w)/B(t_w, t_w)$, where $B(t, t_w) = \sum_i [\langle n_i(t + t_w) n_i(t_w)\rangle - \langle n_i(t + t_w)\rangle\langle n_i(t_w)\rangle]$. In the high T_Γ region, $C(t, t_w)$ has a time translation invariant (TTI) behavior, i.e., $C(t, t_w) = C(t)$. Asymptotically $C(t)$ can be well fitted by stretched exponentials: $C(t) = C_0 \exp[-(t/\tau)^\beta]$. The exponent β becomes significantly lower than 1 at low amplitudes. The above fit defines the relaxation time $\tau(T_\Gamma)$:the growth of τ by decreasing T_Γ is well approximated by an Arrhenius or Vogel-Tamman-Fulcher law (as early found in [7,10]), resembling the slowing down of glass formers close to the glass transition, a result also recently experimentally reported in granular media [5,6]: $\tau \simeq \tau_0 \exp\left[E_0/(T_\Gamma - T_\Gamma^K)\right]$. The divergence point, T_Γ^K (which in simulations is difficult to precisely locate; here it is between 0 and 1), of τ is interpreted as the numerical location of the point of dynamical arrest of the system, where an "ideal" transition to a glassy phase occurs. By quenching the system at low values of T_Γ, the TTI character of relaxation is lost and logarithmic aging behaviors are found. For slow quenches the hard spheres model is able, anyway, to attain its crystal phase. The precise nature of the "glassy" region is, however, very difficult to be numerically determined, so in the next section an analytic approach is presented.

An other dynamical quantity of interest is grains mean square displacement, $R^2(t)$, defining for $t \to \infty$ the diffusion coefficient $D(\rho) = R^2(t)/t$. As much as in glass formers, D goes to zero around a high density ρ_c, signaling a localization transition in which particles are confined in local cages and the macroscopic diffusion-like processes are suppressed [7]. This phenomenon may also be described in a different way: ρ_c is the density above which it becomes impossible to obtain a macroscopic rearrangement of particles positions without increasing the system volume, i.e., the density at which macroscopic shear in the system is impossible without dilatancy. This remark outlines the close correspondence with the phenomenon of the Reynolds transition in granular media [7].

5. A mean field study of hard spheres under gravity on a random graph

We have discussed up to here schematic models for granular media shown to be well described by Edwards' assumption. Since the exact calculation of Edwards partition function, Z, for the above model of monodisperse hard spheres on a cubic lattice is hardly feasible, we now discuss a mean field theory [35]. In an approximation á la Bethe-

Peierls, we consider a random graph version of such a lattice, sketched in Fig. 6: more specifically we consider a 3D lattice box with H horizontal layers (i.e., $z \in \{1, ..., H\}$) occupied by hard spheres; each layer is a random graph of given connectivity, $k-1$ (we take $k = 4$); each site in layer z is also connected to its homologous site in $z-1$ and $z+1$ (the total connectivity is thus $k+1$). The Hamiltonian is the one of Eq.(5) plus a chemical potential term to control the overall density. Hard Core repulsion prevents two connected sites to be occupied at the same time. In the present lattice model we adopt a simple definition of "mechanical stability": a grain is "stable" if it has a grain underneath. The operator Π_r has thus a simple expression: $\Pi_r = \lim_{K\to\infty} \exp\{-K\mathcal{H}_{Edw}\}$ where $\mathcal{H}_{Edw} = \sum_i \delta_{n_i(z),1}\delta_{n_i(z-1),0}\delta_{n_i(z-2),0}$ (for clarity, we have shown the z dependence in $n_i(z)$).

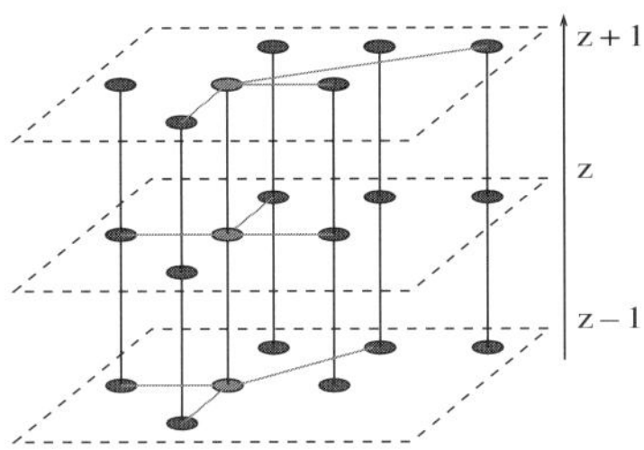

Figure 6. In our mean field approximation, the hard spheres describing the system grains are located on a Bethe lattice, sketched in the figure, where each horizontal layer is a random graph of given connectivity. Homologous sites on neighboring layers are also linked and the overall connectivity, c, of the vertices is $c \equiv k+1 = 5$.

By using the "cavity method" [2], the phase diagram is found [35]. At low N_s or high T_{conf} a fluid-like phase is found, characterized by a homogeneous Replica Symmetric (RS) solution, in which only one pure state exists and the local fields are the same for all the sites of the lattice (translational invariance). For a given N_s, by lowering T_{conf} (see Figs. 7 and 8), a phase transition to a crystal phase (an RS solution with no space translation invariance) is found at T_m. Notice that the fluid phase still exists below T_m as a metastable phase corresponding to a supercooled fluid found when crystallization is avoided. Within the one-step replica symmetry breaking (1RSB) ansatz of the cavity method [2], a non trivial solution appears for the first time at a given temperature $T_D(N_s)$, signaling the existence of an exponentially high number of pure states. In mean field theory T_D is interpreted as the location of a purely dynamical transition as in mode-coupling theory, but in real systems it might correspond just to a crossover in the dynamics (see [26,13,36] and Ref.s therein). The 1RSB solution becomes stable at a lower point T_K, where a thermodynamic transition from the supercooled fluid to a 1RSB glassy phase takes place (see Fig. 7) in a scenario á la Kauzmann with a vanishing complexity of pure states (which stays finite for $T_K < T < T_D$).

The results of these calculations, summarized in the phase diagram of Fig. 7, are further

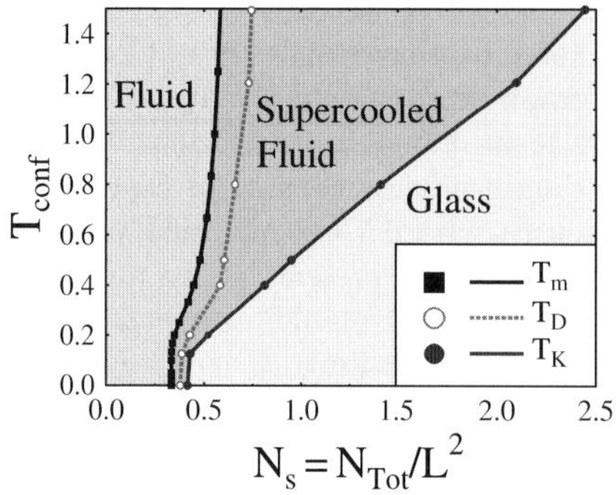

Figure 7. The system mean field phase diagram is plotted in the plane of its two control parameters (T_{conf}, N_s): T_{conf} is Edwards' "configurational temperature" and N_s the average number of grains per unit surface in the box. At low N_s or high T_{conf}, the system is found in a fluid phase. The fluid forms a crystal below a melting transition line $T_m(N_s)$. When crystallization is avoided, the "supercooled" (i.e., metastable) fluid has a thermodynamic phase transition, at a point $T_K(N_s)$, to a Replica Symmetry Breaking "glassy" phase with the same structure found in mean field theory of glass formers. In between $T_m(N_s)$ and $T_K(N_s)$ a dynamical freezing point, $T_D(N_s)$, is located, where the system characteristic time scales diverge.

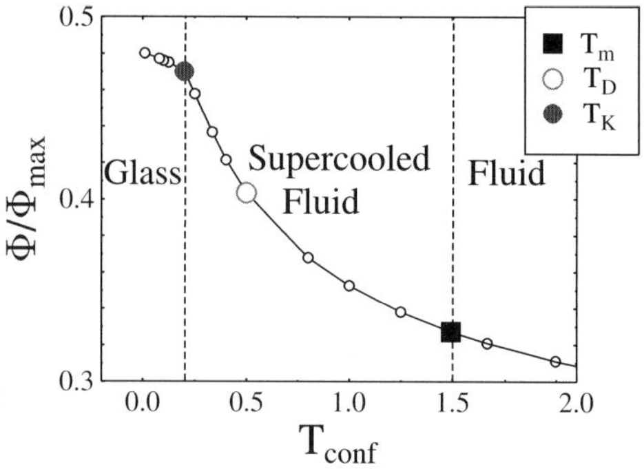

Figure 8. For a system with a given number of grains (i.e., a given N_s), the overall number density, $\Phi \equiv N_s/2\langle z\rangle$ ($\langle z\rangle$ is the average height), calculated in mean field approximation is plotted as a function of T_{conf}; $\Phi(T_{conf})$ has a shape very similar to the one observed in the "reversible regime" of tap experiments and MC simulations of the cubic lattice model for $\Phi(T_\Gamma)$. The location of the glass transition, T_K (filled circle), corresponds to a cusp in the function $\Phi(T_{conf})$. The passage from the fluid to supercooled fluid is T_m (filled square). The dynamical crossover point T_D is found around the flex of $\Phi(T_{conf})$ and well corresponds to the position of a characteristic shaking amplitude Γ^* found in experiments and simulations where the "irreversible" and "reversible" regimes approximately meet.

illustrated in Fig. 8: in a system with a given number of grains (i.e., a given N_s), the overall number density, Φ, is plotted as a function of T_{conf} (here by definition $\Phi \equiv N_s/2\langle z\rangle$, where $\langle z\rangle$ is the average height). The shown curve, $\Phi(T_{conf})$, is the equilibrium function here calculated. It has a shape very similar to the one observed in tap experiments [4,6], or in MC simulations on the cubic lattice (see also [7]), where the density is plotted as a function of the shaking amplitude Γ (along the so called "reversible branch"). In particular, a comparison of our mean field results with simulations of the 3D model of Hard Spheres under the tap dynamics shows a very good agreement.

Summarizing, in the present mean field scenario of a granular medium with N_s particles per surface, in general, at high T_{conf} (i.e. high shaking amplitudes) a fluid phase is located (see Fig. 7). By lowering T_{conf}, a phase transition to a crystal phase is found at T_m. However, when crystallization is avoided, the fluid phase still exists below T_m as a metastable phase corresponding to a supercooled fluid. At a lower point, T_D, an exponentially high number of new metastable states appears, interpreted, at a mean field level, as the location of a purely dynamical transition, which in real system is thought to correspond just to a dynamical crossover. Finally, at a even lower point, T_K, the supercooled fluid has a genuinely thermodynamics discontinuous phase transition to glassy state. The structure of the glass transition of the present model for granular media, obtained in the framework of Edwards' theory, is the same found in the glass transition of the p-spin glass and in other mean field models for glass formers [26,13].

6. A hard-sphere binary mixture under gravity

In order to study segregation and to test Edwards proposal to a more complicate system, here we consider a hard-sphere binary system made of two species 1 (small) and 2 (large) with grain diameters a_0 and $\sqrt{2}a_0$, under gravity on a cubic lattice of spacing $a_0 = 1$. We set the units such that the two kinds of grain have masses $m_1 = 1$ and $m_2 = 2m_1$, and gravity acceleration is $g = 1$. The hard core potential $\mathcal{H}_{HC}$ is such that two large nearest neighbor particles cannot overlap. This implies that only couples of small particles can be nearest neighbors on the lattice. The system overall Hamiltonian is:

$$\mathcal{H} = \mathcal{H}_{HC} + m_1 g H_1 + m_2 g H_2, \tag{7}$$

where $H_1 = \sum_i^{(1)} z_i$ and $H_2 = \sum_i^{(2)} z_i$, the height of site i is z_i and the two sums are over all particles of species 1 and 2, respectively. In the above units, the gravitational energies in a given configuration are thus $E_1 = H_1$ and $E_2 = 2H_2$.

As before, grains are confined in a box of linear size L between with periodic boundary conditions in the horizontal directions and initially prepared in a random loose stable pack. Under the tap dynamics the system approaches a stationary state for each value of the tap parameters T_Γ and τ_0 used. In Fig. 9, we plot as function of T_Γ (for several values of τ_0) the asymptotic value of the *vertical* segregation parameter, i.e., the difference of the average heights of the small and large grains at stationarity, $\Delta h(T_\Gamma, \tau_0) \equiv h_1 - h_2$. Here h_1 and h_2 are the average of H_1/N_1 and H_2/N_2 over the tap dynamics at stationarity.

The results given in the main panel of Fig. 9 apparently show that T_Γ is not a right thermodynamic parameter, since sequences of taps with different τ_0 give different values for the system observables. However, if the stationary states corresponding to different tap dynamics (i.e., different T_Γ and τ_0) are indeed characterized by a single thermodynamic parameter, as in the monodisperse case above, the curves of Fig. 9 should collapse onto a universal master function when $\Delta h(T_\Gamma, \tau_0)$ is parametrically plotted as function of an other macroscopic observable such as the average energy, $e(T_\Gamma, \tau_0) = (E_1 + E_2)/N$ (N is the total number of particles). This collapse of data is not observed here, as it is apparent in the inset of Fig. 9. We found, instead, [10] that two macroscopic quantities can be sufficient to characterize uniquely the stationary state of the system. These two quantities are, for instance, the energy e and the height difference Δh. Of course since $e = ah_1 + 2bh_2$ (where $a = N_1/N$ and $b = N_2/N$) and $\Delta h = h_1 - h_2$, we can also choose h_1 and h_2 to characterize the stationary state. Namely, we found that a generic macroscopic quantity A, averaged over the tap dynamics in the stationary state, is only dependent on h_1 and h_2, i.e., $A = A(h_1, h_2)$. We have checked that this is the case for several independent observables, such as the number of contacts between large particles, N_c, the density of

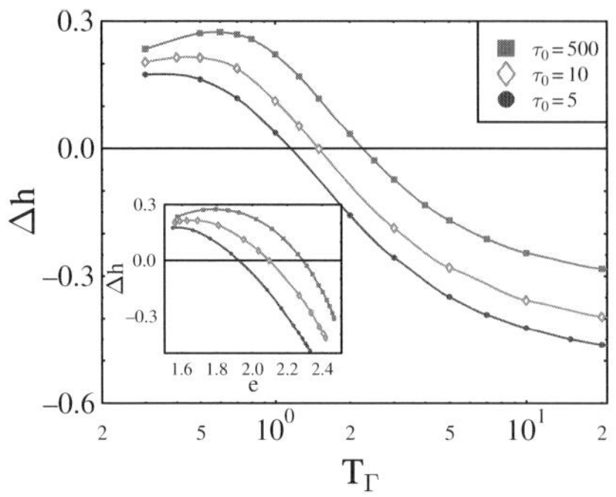

Figure 9. **Main frame** The difference of the average heights of small and large grains, $\Delta h = h_1 - h_2$, measured at stationarity in the binary hard spheres mixture under gravity, is plotted as a function of tap amplitude, T_Γ (in units mga_0). The three sets of points correspond to the shown tap durations, τ_0. At high T_Γ larger grains are found above the smaller, i.e, $\Delta h < 0$, as in the Brazil nut effect (BNE). Below a $T_\Gamma^*(\tau_0)$ the opposite is found (Reverse Brazil nut effect, RBNE). **Inset** The Δh data of the main frame are plotted as a function of the corresponding average energy, e. The three sets of data do not collapse, as before, onto a single master function.

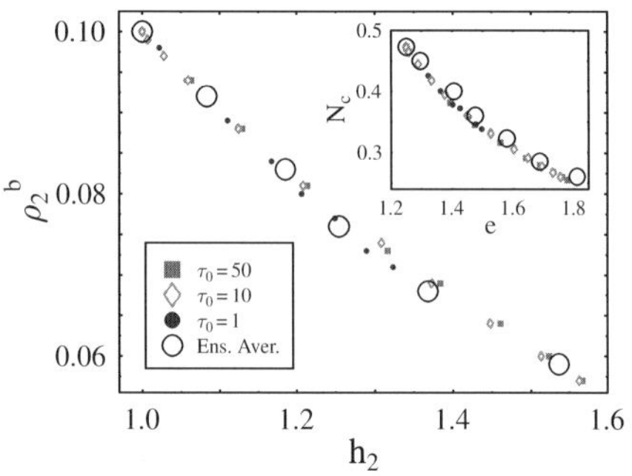

Figure 10. **Main frame** The average density of large grains on the box bottom layer, ρ_2^b, measured at stationarity for different T_Γ and τ_0, scale almost on a single master function when plotted as a function of the large grains height, h_2. **Upper inset** The average number of contacts between large grains per particle, N_c, obtained for different T_Γ and τ_0, scale on a single master function when plotted as a function of the system energy, e.

small and large particles on the bottom layer, ρ_1^b and ρ_2^b, and others. In particular, as shown in Fig. 10, we find with good approximation that: $N_c \simeq N_c(e) = N_c(ah_1 + bh_2)$, $\rho_2^b \simeq \rho_2^b(h_2)$, $\rho_1^b \simeq \rho_1^b(h_1)$. Therefore we need both h_1 and h_2 to characterize unambiguously the state of the system; namely all the observables assume the same values in a stationary state characterized by the same values of h_1 and h_2, independently on the previous history (i.e., in our case independently on the particular tapping parameters T_Γ and τ_0).

These findings imply that an extension of Edwards' approach is required, where at least *two* thermodynamic parameters have to be included [10]. As before, this can be obtained by assuming that the microscopic distribution is given by the principle of maximum entropy with the constraint that the average gravitational energies of the two species $E_1 = \sum_r P_r E_{1r}$ and $E_2 = \sum_r P_r E_{2r}$ are independently fixed. This gives *two* Lagrange multipliers:

$$\beta_1 = \frac{\partial \ln \Omega_{IS}(E_1, E_2)}{\partial E_1} \qquad \beta_2 = \frac{\partial \ln \Omega_{IS}(E_1, E_2)}{\partial E_2}$$

where $\Omega_{IS}(E_1, E_2)$ is the number of inherent states with E_1,E_2.

The hypothesis that the ensemble distribution at stationarity is the above can be tested as we have already previously shown. We have to check that the time average of any quantity, $A(h_1, h_2)$, as recorded during the taps sequences in a stationary state characterized by given values h_1 and h_2, coincides with the ensemble average, $\langle A\rangle(h_1, h_2)$, over the generalized version of distribution Eq.(2). To this aim, we have calculated the ensemble averages $\langle N_c\rangle$, $\langle\rho_2^b\rangle$, $\langle\rho_1^b\rangle$ for different values of β_1 and β_2; we have expressed parametrically $\langle N_c\rangle$, $\langle\rho_2^b\rangle$, $\langle\rho_1^b\rangle$, as function of the average of h_1 and h_2, and compared them with the corresponding quantities, N_c, ρ_1^b and ρ_2^b, averaged over the tap dynamics. The two sets of data are plotted in Fig. 10 showing a good agreement (notice, there are no adjustable parameters).

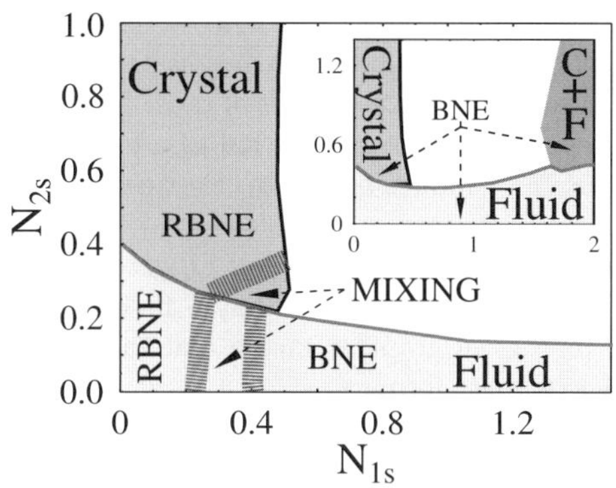

Figure 11. Mean field phase diagram of the binary mixture in the plane (N_{1s}, N_{2s}) (particles densities per unit surface of species 1 and 2) for $m_1\beta_1 = 0.8$ and $m_2\beta_2 = 1.25$ (**main frame**). A fluid and a crystal phase are present (glassy and metastable phases are not shown here). Size segregation is found as a form of phase separation when the system is prepared out of these phases. Within a given phase, gravity drives vertical mixing/segregation phenomena (the BNE, RBNE and MIXING regions) separated by smooth crossovers (shaded areas). **Inset** Phase diagram for $m_1\beta_1 = 1.25$ and $m_2\beta_2 = 0.8$. Here also a new phase, with strong BNE, appears where a fluid, rich in small grains, is found beneath a crystal rich in large ones ("C+F" region). For $m_1\beta_1 << m_2\beta_2$, strong RBNE can be found, with the fluid above the crystal.

In [37] the mean field approximation of Sec. 5 is applied to the present binary mixture, giving a precise interpretation of segregation phenomena observed in the model of Eq.(7), as shown in Fig.11. In a nutshell, by changing N_{1s}, N_{2s} (i.e., densities per unit surface of species 1 and 2), β_1 and β_2 (i.e, their configurational temperatures), or masses or sizes ratios, the system exhibits true phase transitions from fluid to crystal phases. As much as in thermal media, this induces segregation effects associated to phase separation phenomena, with the formation of coexisting phases rich in small or large grains, as experimentally observed in [22]. Within a given phase, the presence of gravity can also drive a form of "vertical" segregation, where large grains are found on average above (the well known BNE) or below small grains (RBNE [20]), but this is usually not directly associated to phase transitions.

7. Conclusions

Summarizing, within the schematic framework of lattice models, we have shown, by Monte Carlo simulations of "tap" dynamics, that Edwards' approach to granular media at rest appears to be well grounded. The system stationary states are indeed independent on the sample history as in a "thermodynamics" system, and can be described in terms of a distribution function characterized by a few control parameters (such as configurational temperatures). By use of Edwards approach, we have derived, by analytical calculations at a mean field level, the phase diagram of these systems. In particular, we discovered that "jamming" corresponds to a phase transition from a "fluid" to a "glassy" phase, observed when crystallization is avoided. Interestingly, the nature of such a "glassy" phase turns out to be the same found in mean field models for glass formers. In the same framework, we have also briefly discussed segregation patterns observed in a hard sphere binary model under gravity subject to sequences of taps. Here, the presence of fluid-crystal phase transitions in the system drives segregation as a form of phase separation. Within a given phase, gravity can also induce a kind of "vertical" segregation, not associated to phase transitions.

In practice, even though the general validity of Edwards approach to "frozen" systems has just begun to be assessed, it turns out that a first reference framework is emerging to understand the physics of granular media and their deep connections with other "jamming" systems such as glass formers.

REFERENCES

1. S. F. Edwards and R. B. S. Oakeshott, Physica A 157 (1989) 1080. A. Mehta and S. F. Edwards, Physica A 157 (1989) 1091; S.F. Edwards, in *"Disorder in Condensed Matter Physics"* p. 148, Oxford Science Pubs (1991); and in *Granular Matter: an interdisciplinary approach*, (Springer-Verlag, New York, 1994), A. Mehta ed.
2. M. Mézard and G. Parisi, Eur. Phys. J. B 20 (2001) 217; M. Mézard and G. Parisi, J. Stat. Phys. 111 (2003) 1.
3. H. M. Jaeger, S. R. Nagel and R. P. Behringer, Rev. Mod. Phys. 68 (1996) 1259.
4. J. B. Knight, C. G. Fandrich, C. N. Lau, H. M. Jaeger and S. R. Nagel, Phys. Rev. E 51 (1995) 3957; E. R. Nowak, J. B. Knight, E. Ben-Naim, H. M. Jaeger and S. R. Nagel, Phys. Rev. E 57 (1998) 1971; E. R. Nowak, J B. Knight, M. Povinelli, H. M. Jaeger and S. R. Nagel, Powder Technology 94 (1997) 79.
5. G. D'Anna and G. Gremaud, Nature 413 (2001) 407.
6. P. Philippe and D. Bideau, Europhys. Lett. 60 (2002) 677.
7. M. Nicodemi, A. Coniglio and H.J. Herrmann, Phys. Rev. E 55 (1997) 3962; M. Nicodemi, A. Coniglio and H.J. Herrmann, J. Phys. A 30 (1997) L379.
8. A. J. Liu and S. R. Nagel, Nature 396 (1998) 21.
9. C. S. O'Hern, S. A. Langer, A. J. Liu and S. R. Nagel, Phys. Rev. Lett. 86 (2001) 111. C. S. O'Hern, L. E. Silbert, A. J. Liu and S. R. Nagel, Phys. Rev. E 68, 011306 (2003).
10. A. Fierro, M. Nicodemi and A. Coniglio, Europhys. Lett. 59 (2002) 642; Phys. Rev. E 66 (2002) 061301; Europhys. Lett. 60 (2002) 684.
11. F. Sciortino, W. Kob and P. Tartaglia, Phys. Rev. Lett. 83 (1999) 3214.

12. W. Kob, F. Sciortino and P. Tartaglia, Europhys. Lett. 49 (2000) 590. F. Sciortino and P. Tartaglia, Phys. Rev. Lett. 86 (2001) 107.
13. G. Biroli and M. Mézard, Phys. Rev. Lett. 88 (2002) 025501.
14. H. A. Makse and J. Kurchan, Nature 415 (2002) 614.
15. J. M. Ottino and D. V. Khakhar, Ann. Rev. Fluid Mech. 32 (2000) 55. J. Bridgewater, Chem. Eng. Sci. 50 (1994) 4081.
16. T. Rosato, F. Prinze, K. J. Standburg and R. Swendsen, Phys. Rev. Lett. 58 (1987) 1038.
17. J. Bridgewater, Powder Technol. 15 (1976) 215; J. C. Williams, Powder Technol. 15 (1976) 245.
18. J. Duran, J. Rajchenbach and E. Clement, Phys. Rev. Lett. 70 (1993) 2431.
19. J. Knight, H. Jaeger and S. Nagel, Phys. Rev. Lett. 70 (1993) 3728.
20. D. C. Hong, P. V. Quinn and S. Luding, Phys. Rev. Lett. 86 (2001) 3423; J. A. Both and D. C. Hong, Phys. Rev. Lett. 88 (2002) 124301.
21. T. Shinbrot and F. Muzzio, Phys. Rev. Lett. 81 (1988) 4365.
22. K. M. Hill and J. Kakalios, Phys. Rev. E **49**, R3610 (1994). P.M. Reis and T. Mullin Phys. Rev. Lett. **89**, 244301 (2002). S. Aumaitre, T. Schnautz, C.A. Kruelle, and I. Rehberg, Phys. Rev. Lett. **90**, 114302 (2003).
23. F. H. Stillinger and T. A. Weber, Phys. Rev. A 25 (1982) 978; S. Sastry, P. G. Debenedetti and F. H. Stillinger, Nature 393 (1998) 554.
24. B. Coluzzi, G. Parisi and P. Verrocchio, Phys. Rev. Lett. 84 (2000) 306.
25. R. Monasson, Phys. Rev. Lett. 75 (1995) 2847; Th. M. Nieuwenhuizen, Phys. Rev. E 61 (2000) 267; S. Franz and M. A. Virasoro, J. Phys. A 33 (2000) 891; A. Crisanti and F. Ritort, J. Chem. Phys. 113 (2000) 10615.
26. L. Cugliandolo and J. Kurchan, Phys. Rev. Lett. 71 173 1993. J. Kurchan, in *"Jamming and Rheology: Constrained Dynamics on Microscopic and Macroscopic Scales"* A.J. Liu and S.R. Nagel Eds., Taylor and Francis, London, 2001.
27. M. Nicodemi, Phys. Rev. Lett. 82 (1999) 3734.
28. A. Barrat *et al.*, Phys. Rev. Lett. 85 (2000) 5034.
29. A. Coniglio and M. Nicodemi, Physica A 296 (2001) 451.
30. J. J. Brey, A. Prados and B. Sánchez-Rey, Physica A 275 (2000) 310.
31. D. S. Dean and A. Lefèvre, Phys. Rev. Lett. 86 (2001) 5639.
32. J. Berg, S. Franz and M. Sellitto, Eur. Phys. J. B 26 (2002) 349.
33. J. Berg and A. Mehta, Europhys. Lett. 56 (2001) 784.
34. A. Coniglio, A. de Candia, A. Fierro and M. Nicodemi, Jour. Phys.: Cond. Mat. 11 (1999) A167.
35. A. Coniglio, A. de Candia, A. Fierro, M. Nicodemi and M. Tarzia, *cond-mat/0309612.*
36. C. Toninelli, G. Biroli and D.S. Fisher, *cond-mat/0306746.*
37. A. Coniglio, A. Fierro, M. Nicodemi and M. Tarzia, in preparation (2003).

Unifying Concepts in Granular Media and Glasses
A. Coniglio, A. Fierro, H.J. Herrmann and M. Nicodemi (editors)

Granular compaction

Daniel Bideau, Pierre Philippe, Philippe Ribière and Patrick Richard [a]

[a]Groupe Matière Condensée et Matériaux, UMR CNRS 6626, F-35042 Rennes cedex

We report experimental results on granular compaction under consecutive vertical taps. The evolution of the mean volume fraction presents a slow densification until a final steady state and follows a KWW law. This result does not depend on the shape of the grains. The link between compaction and convection is also studied. Finally, a local analysis of the packing is discussed.

1. Introduction

When continuously submitted to weak mechanical taps, a granular medium slowly compacts. This compaction is essentially ruled by geometrical constraints between grains ; the thermal energy k_BT does not play any role, because it is negligible compared to the gravitational energy of a grain. Each packing is for this reason in a metastable configuration, which remains stable as long as there is no external - mechanical - perturbation. In this framework, compaction of grains under mechanical driving is a good practical way to study the succession of jumps for a metastable equilibrium to another one. The first experiments in this field were done in Chicago [1]. Starting from a loose packing of beads confined in a tube, a succession of vertical taps of controlled acceleration induces a progressive and very slow compaction of the system : after more than 10 000 taps, a steady state is still not reached. An empirical law in the inverse of the logarithm of the number of taps (in other words the time) was proposed to describe the evolution of the volume fraction. These results have motivated many theoretical and numerical works, most of them dealing with the concept of free volume or geometrical constraint [2–9]. Some of them underscore structural aging effects, as currently noticed in glassy systems. We present here new compaction experiments in what we believe to be more general conditions : the experiments is geometrically different, we use different types of grains...

2. Experimental set-up

Nearly monodisperse glass beads and in a second type of experiment rice grains are filled to a height of about 10 cm in a glass cylinder (diameter D = 10 cm). We are able to prepare the sample in a reproducible loose packing state where the initial mean packing fraction $\langle \Phi_0 \rangle \approx 0.58$. This container is placed on the plate connected to an electromagnetic exciter (LDS V406) which induces a vertical displacement to the plate. The container is in this way shaken at regular intervals ($\Delta t = 0.5$s) by vertical taps. Each tap is controlled by an entire cycle of sine wave at a fixed frequency $f = 30$ Hz. The resulting motion of the whole system, monitored by an accelerometer at the bottom of the container, is however more complicated than a simple sine wave: at first, the system undergoes a positive acceleration followed by a negative peak with a minimum equal to $-\gamma_{max}$. After the applied voltage stops, the system relaxes to its normal repose position. A scheme of the experimental set-up is shown figure 1. An experiment is realized as follows. We start

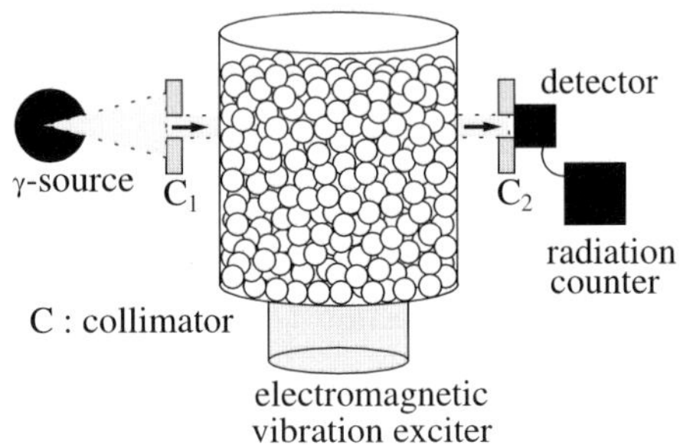

Figure 1. Scheme of the experimental set-up

from a reproducible loose packing ($\Phi_0 = 0,583 \pm 0,003$), obtained by filling the container through a grid. Another grid initially placed at the bottom of the container is then pulled up in order to destructure the packing. Then, sequences of 10^4 to 10^6 taps are carried out with a given acceleration $\Gamma = \gamma_{max}/g$, where g is the gravitational acceleration. The average volume fraction in the bulk, $\langle\Phi\rangle$, is measured from the transmission ratio of a γ-ray beam through the packing. This γ-detection permits to measure not only $\langle\Phi\rangle$, but also the vertical density profile $\Phi(z)$ of the packing the plate supporting the container has 3 degrees of freedom in translation and 1 in rotation. The value of Φ is deduced from the transmission ratio T of a horizontal collimated γ-beam through the packing. $T = A/A_0$ where A and A_0 are respectively the activities counted by the detector with or without the beads in the container. Equation 1, derived from the Beer law for absorption gives an estimation of the volume fraction in the probe zone :

$$\Phi \approx \ln(T)/\mu D. \quad (1)$$

Here μ is the absorption coefficient of the beads, estimated experimentally. The collimated γ-beam is nearly cylindrical with a diameter of 10 mm and intercepts perpendicularly the vertical axis of the cylindrical container. An acquisition-time of 60 s was found to be a good compromise between the intrinsic uncertainty of a radioactive beam and the total duration of an experiment. We then achieve a precision $\Delta\Phi = \pm 0.003$. With the aim of limiting the duration of the experiments (from 20 hours to more than a month according to the total number of taps) and of avoiding redundant information due to the very slow evolution of the system, the measurements are spaced out in time, on a logarithmic scale, with 2 measures of profile and 50 measures of $\langle\Phi\rangle$ per decade (except 10 for the first decade).

3. Experimental results on 1 mm spheres

Several compaction experiments have been carried out for different values of the tapping strength Γ between 0 and 6. Part of the results are presented in figure 2 and show the evolution of the mean volume fraction $\langle\Phi\rangle$ as function of the number of taping and for a few values of Γ. The curves are averaged on 2 or 3 experiments : t is in fact the number of taps, and the observed dynamics is the result of a succession of metastable states induced by the taps. Because of the very slow evolution, we have used a logarithmic time scale, where $\langle\Phi\rangle$ is plotted versus $\log(t+1)$. The typical evolution is a slow compaction of the packing characterized by an increase of $\langle\Phi\rangle$; then after a varying relaxation time (depending on Γ), the system finally reaches a steady state. Note that for $\Gamma = 0.96$, the temporal window accessible in our experiment is too small to observe the whole relaxation process. In all our experiments, the volume fraction remains less than the random close packing limit and no order is observed from the glass walls of the container. In comparison with the previous experimental results obtained by the Chicago group, some differences are observed. For the same values of Γ, the dynamics of compaction is faster in our case, particularly for the highest values of Γ, and a final steady state is here definitively observed. This difference leads to a different fit for $\langle\Phi\rangle = f(t)$: the best fit is not the $1/\log(t)$ proposed in [1],

$$\Phi(t) = \Phi_{ss} - \frac{(\Phi_{ss} - \Phi_0)}{1 + B\ln(1 + t/\tau)},$$

but is a stretched exponential function [10], first proposed by Kohlrausch in 1854 [11] and far later popularized by Williams and Watts [12]. This fit can be expressed as :

$$\phi(t) = \phi_{ss} - (\phi_{ss} - \phi_0)\exp\left(-\left(\frac{t}{\tau}\right)^{\beta}\right). \tag{2}$$

Such an expression is now frequently applied to a large range of relaxations in disordered thermal systems like glasses (see for instance ref [10]). Examples of this fit are presented in figure 2 where solid lines are stretched exponential laws. To quantify the influence of Γ on the dynamics of compaction, we have made an experimental evaluation of the relaxation time. First we use two different initial packings : a loose one with a packing fraction equal to 0.583 ± 0.003 and a more compact one with $\Phi = 0.632 \pm 0.003$. These packings are submitted to taps with the same Γ. The first one densifies whereas the other dilates and both them reach the same steady-state. We can then estimate a time of convergence τ_{conv} which can be considered as a characteristic relaxation time, or a memory duration that is the number of taps required for a packing to forget its initial configuration. If we plot this time τ_{conv} versus the acceleration Γ, we find an Arrhenius law (see figure 3) of the type of equation 3 :

$$\tau_{conv}(\Gamma) = \tau_0 \exp\left(\frac{\Gamma_0}{\Gamma}\right), \tag{3}$$

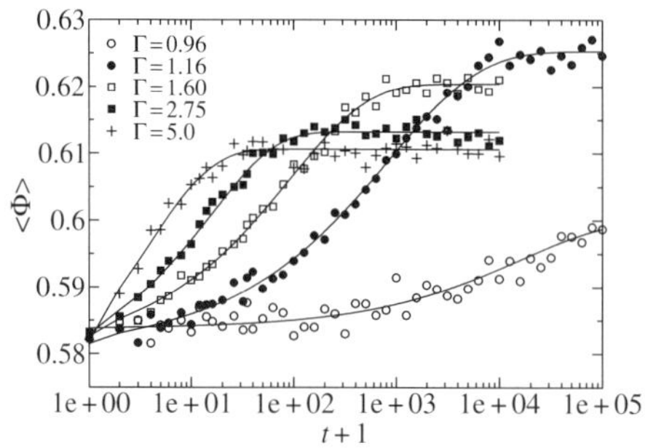

Figure 2. Temporal evolution of the mean volume fraction $\langle\Phi\rangle$ for different tapping intensities. To make the curves easier to examine, only 25% of the points are plotted.

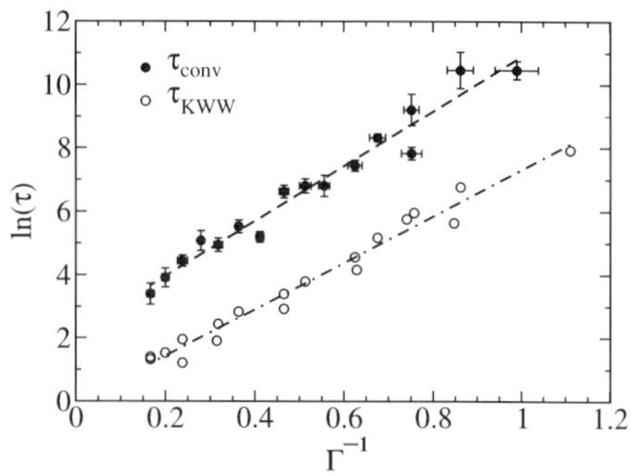

Figure 3. Experimental estimation of the relaxation time on the tap acceleration.

where $\ln \tau_0 \approx 2.2$ and $\Gamma_0 \approx 8.5$. Note that the characteristic time extracted from KWW fit also follows an Arrhenius law (see figure 3, open circles).

Because the friction on the walls of the container, the threshold Γ^* above which there is a collective lift-off of the packing is slightly greater than 1 and is of the order of 1.2. The dynamics is different for $\Gamma < \Gamma^*$ and $\Gamma > \Gamma^*$, as shown figure 4. For $\Gamma < \Gamma^*$, compaction is still observed but with a slower dynamics. For most of these experiments, we were only able to observe the beginning of the relaxation and it is difficult to be sure that a stretched exponential law still describes the whole compaction dynamics. If we postulate the validity of the KWW law, we thus obtain the characteristic time τ_{KWW}. In this case Φ_{ss} cannot be directly measured from the final value of the volume fraction. But using the dependence of Φ_{ss} on Γ presented in the inset of figure 4, we can extrapolate Φ_{ss} between 0.625 and 0.63. The new behavior obtained for the smallest values of Γ is still compatible with an Arrhenius law but with very different values of the two parameters of equation 3 : $\Gamma_0 \approx 24$ and $\ln \tau_0 \approx -14$. Another difference with the Chicago's experiment is that we observe convection during compaction. Two different regimes are observed according to the value of Γ, with a cross-over Γ_c near $\Gamma = 2$. It corresponds to a significant change in the dependence of Φ_{ss} on Γ (see the inset of figure 4). Above Γ_c, the free surface heaps up moderately and finally takes on a flat conical shape, probably brought about by nearly toroïdal convection roll which is the expected form of the convective flow in a cylindrical with high enough accelerations [14]. Below Γ_c, the free surface progressively slopes along a preferred orientation which indicates a spontaneous breaking of the cylindrical symmetry on behalf of a plane symmetry (see figure 5). The steady state slope of the inclined plane tends to increase when Γ decreases, but the time for this to happen becomes extremely long. These types of free surface instabilities have been observed previously [15]. As shown in figure 6, the convective flow observed through the displacement of dyed tracers near the wall is asymmetrical with a fastest zone below the bottom of the inclined free

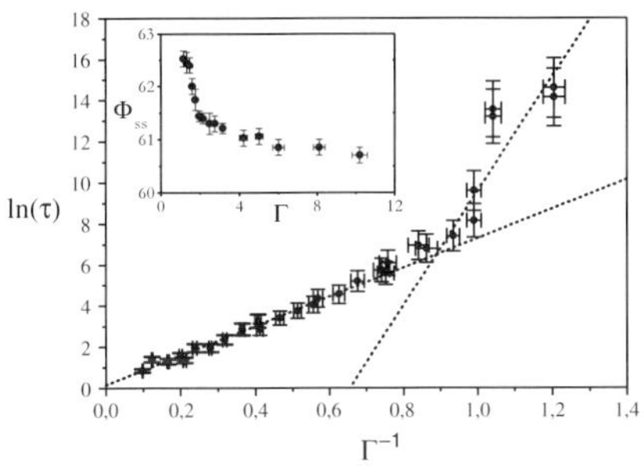

Figure 4. Dependence of the compaction time τ_{conv} on the inverse of the acceleration Γ. The dotted lines represent the two Arrhenius laws. Inset: variation of the final packing fraction Φ_{ss} with Γ in cases where a steady state is actually reached.

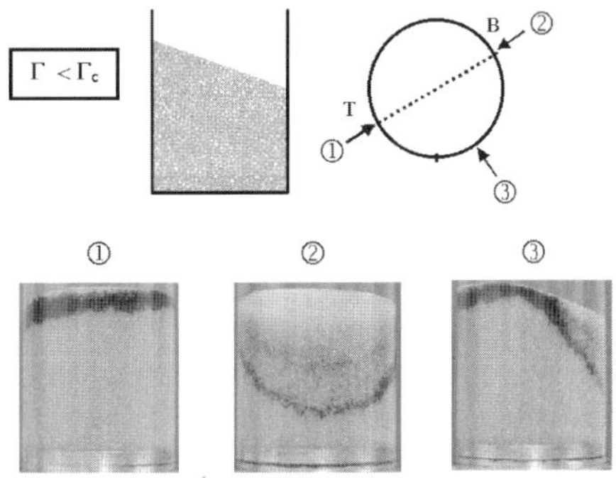

Figure 5. Convection at low accelerations. The drawings are the front view of the cylindrical container, with the free surface, and the above view, with the symmetrical plane in dotted line (T and B denote the top and the bottom of the surface). The photographs correspond to three different views indicated on the upper drawing.

surface. The same experiments above Γ_c show a much less asymmetrical flow. Briefly, we have verified that an unavoidable and very slight bias (less than few tenth of degrees) inherent to the set-up, gives rise to an asymmetrical mass flow and to an inclination of the free surface which, at low accelerations and for long times, cannot be counterbalanced by any surface avalanche induced by the taps. We have evaluated the convection strength through the displacement of a ring of dyed tracers initially placed at the periphery of the free surface. The downward velocity near the wall is deduced from the maximum cumulative displacement of the tracers. It progressively decreases as the packing compacts and finally reaches a steady state value V_{ss} (see figure 6).

But, in the range $\Gamma < \Gamma^*$, a significant displacement of tracers is obtained only after a huge number of taps and the final steady state is absolutely not reached. In this case, it is only possible to roughly estimate the downward velocity by the ratio $V = D_{max}/N$, where D_{max} is the maximum cumulative displacement in bead diameters and N the number of taps. Moreover we have also worked with continuous vibrations instead taping. For each Γ, the convective intensity was estimated by V with an error bar which takes into account the dependence of V on the different choices of N and D_{max}. Although these error bars can be quite large on the linear scale, they remain fairly reasonable on a logarithmic scale used in figure 7. For $\Gamma > \Gamma^*$, the values of V are obtained whether after vibrations or taping, sequences without any appreciable change are observed, and are directly linked to the steady state velocities V_{ss} which are easily measured in this case. It must be pointed out that the velocities V can become extremely small (down to displacement of about one diameter after more than 100 000 cycles) ; extremely small but sufficient to induce spectacular effects in the long time range, such as the inclination of the free surface. As illustrated in figure 6, the dependence on Γ of the convection velocity V is quite consistent with two distinct Arrhenius type behaviors above and below Γ^* :

$$V(\Gamma) = V_0 \exp\left(-\frac{\Gamma_0^V}{\Gamma}\right). \tag{4}$$

Above Γ^*, the parameters are $\Gamma_0^V \approx 8.0$ and $\log V_0 \approx 1.0$. Below there is a sharp deviation from the previous Arrhenius law with a significant slowing down. The corresponding parameters are $\Gamma_0^V \approx 25.4$ and $\log V_0 \approx 15.4$. In all these Arrhenius laws obtained for convection as for compaction dynamics, Γ appears to play an essential role, directly equivalent to a temperature. But it is not systematic and, for instance, equations 3 and 4 are no longer valid in the case of larger beads and must be modified. The very similar dependence of the compaction and the convection on the adimensional acceleration Γ points out a remarkable change of behavior around Γ_c, in the sense of a huge, slowing down of both dynamics as well as a strong coupling also illustrated by a power law with an exponent close to 1 between V and τ (see [16]). As in glasses, where the structural relaxation time or the viscosity is exponentially dependent on the temperature, the compaction time and the inverse of the convection intensity are also exponentially dependent on the strength of the mechanical driving. This result confirms and reinforces the analogy between slightly driven granular media and glass forming materials. But here, the quantity able to characterize correctly the agitation of grains (i.e. the equivalent of a temperature) is not clearly defined and depends directly on the nature of the driving, or in other words on the way of supplying energy to the grains. Above Γ^*, each tap induces a

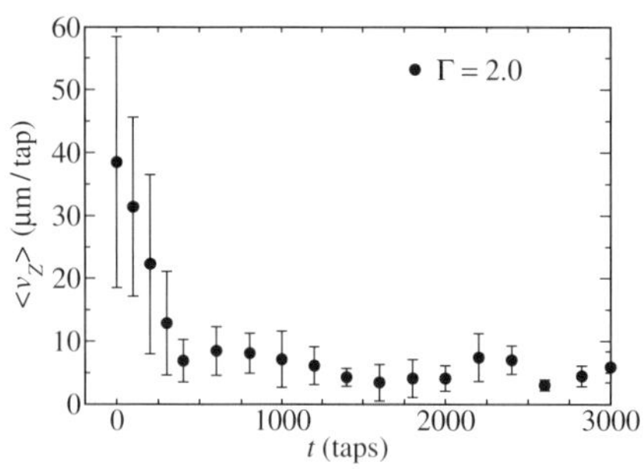

Figure 6. Convection intensity V as the function of time. The system reaches quickly a steady state.

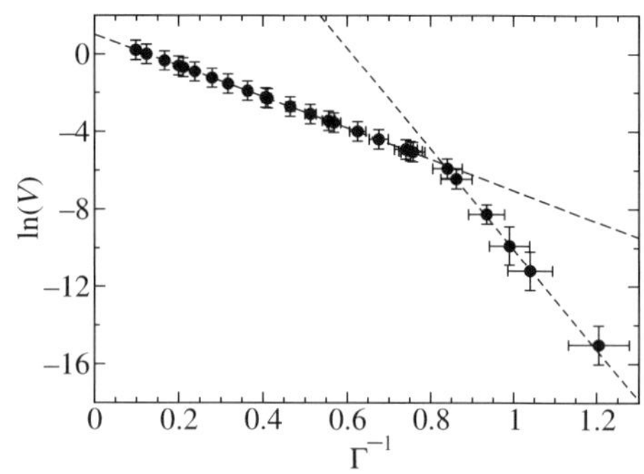

Figure 7. Dependence of the convection intensity V with the inverse of the acceleration parameter Γ. The two Arrhenius laws are plotted in dotted line.

flight of the whole packing as shown in figure 8 (pictures are extracted from a fast camera movie with larger beads to improve visualization without significant qualitative change). During its flight, the packing only dilates very slightly and its maximum altitude is quite small. It means that most of the contacts are maintained during the flight and that the packing keeps the memory of its previous configuration. When the packing crashes

Figure 8. Pictures of a 5 mm beads packing during a tap (Γ = 5.95 and Φ = 30 Hz) : lift-off (t = 0) maximum altitude of the flight (t = 28 ms), beginning of the crash landing (t = 60 ms) intermediate positions (t = 70 ms and 80 ms) and end of the collision (t = 90 ms)

back on the bottom of the container, we observe a shock wave which quickly propagates through the packing in a time of the order of 10 ms and which might densify collectively the grains in each horizontal layer (the horizontal dotted line in figure 8 highlights this overall compaction). The kinetic energy of the packing just before its crash appears to be the amount of energy used to compact the grains during the propagation of the shock wave.

4. Compaction of anisotropic granular media

This part is devoted to the influence of the grain anisotropy on the compaction dynamic. In most previous studies, including those presented in the first sections of this article, the grains used are spheres. This is mainly done to reduce the number of parameters of the physical system. Nevertheless actual grains are far from being perfectly spherical and thus not isotropic. The grain anisotropy is an important parameter since it can drastically modify the grains mobility and thus different results are expected. What results on isotropic beads are still valid for anisotropic grains ? Is the KWW law is still a valid fit ? Moreover compaction experiments with rods were carried out by Villarruel et al. [17] and have shown that initially loose packing evolve to a dense, highly ordered

nematic state. Nevertheless these experiments have been performed with a strong lateral confinement that can induce such kind of order.
As described above, our experimental set-up does not have such confinement and thus we can test if this order is indeed a consequence of the confinement. Two kinds of anisotropic grains are used (see figure 9) : long rice (basmati rice) and short rice (round rice). In order to quantify the grain anisotropy each grain is assimilated to an ellipsoid and the mean ratio between the two axis is measured. We obtain 1.6 for the basmati rice and 1.2 for the round rice. It is worth noting that the size distribution is larger for the round rice.

The method used to build our initial packing is reproducible and gives $\Phi \approx 0.56$ for round rice and $\Phi \approx 0.58$ for basmati rice, which means that the more anisotropic the grains are, the less important the initial packing fraction is. This can be explained by the fact that arches - and thus large pores - form more easily with anisotropic grains. The first observation is that even though the initial packing fraction is lower for the basmati rice than for the round rice, the final and steady state one is larger for this rice : it can be greater than 0.7 for basmati rice whereas it is always less than 0.65 for round rice and than 0.63 for spherical glass. Even if the packing ratio can be really high, we didn't observe, contrary to Villarruel et al. [17] nematic order. This order seems to be indeed due to the strong lateral confinement. Nevertheless, as explained below, we can observe for given values of Γ, order which is not nematic.
The most important results deal with the mechanisms and the dynamic of compaction. Depending on the value of Γ, two kinds of phenomena can be observed : for high Γ convection takes place within the whole media whereas for low values of Γ convection is localized in a part of the packing and is not steady. As it will be shown, compaction mechanisms strongly depend on the convection and thus on Γ. It should be pointed out that we only consider value of Γ larger than the lift-off threshold.

Let us first consider the case of high tap intensity typically $\Gamma > 3$ for basmati rice. The convection takes place quickly in the media and we observe convection rolls at the container wall as showed on figure 10. With regular acquisition of images after each mechanical perturbation, a characteristic time for the evolution of convection rolls, τ_{conv}, can be extracted. Indeed, we can track grain movements and measure how long it takes a grain to revolve around the center of the convection roll. For basmati rice submitted to mechanical perturbations of intensity $\Gamma = 6$, $\tau_{conv} \approx 90$. The evolution of packing fraction during this experiment, is reported on figure 11. The experimental values correspond to black points and we fit the evolution of the packing ratio by two types of law: the Chicago's law and the KWW law. The two curves correctly describe the experimental evolution. However as can be seen on figure 11, the evolution of the Chicago's curve near this steady state is slower than the experimental result. The best fit provides the following values: $\tau = 1280$ and $B = 225$ for Chicago's law and $\tau \approx 19$ and $\beta = 0.31$ for the KWW law. The value of τ is of the same order of magnitude than τ_{conv} and this remains valid for all our experiments done for $\Gamma > 3$.

To go further, we use a high-speed camera (1000 images per second) to analyze the grain movement during a tap. Three phases can be distinguished: during the first one, the rice takes off from the bottom of the container like a solid. During the second phase the rice falls and we observe a dilatation of the media which implies a decompaction. Finally during the third and last phase, a compression wave propagate throughout the media, like

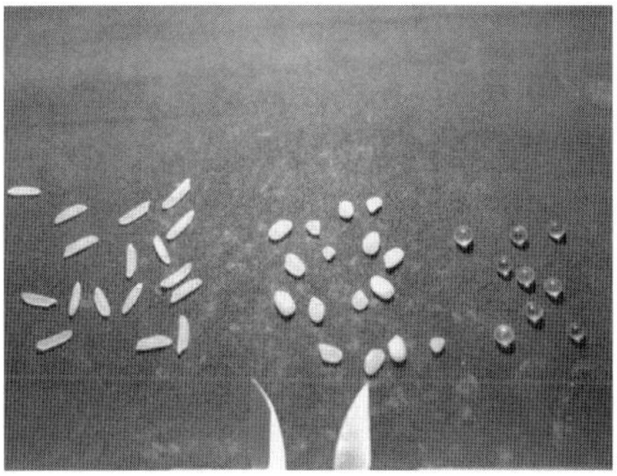

Figure 9. Grains used : basmati rice (left) round rice (center) and glass beads (right)

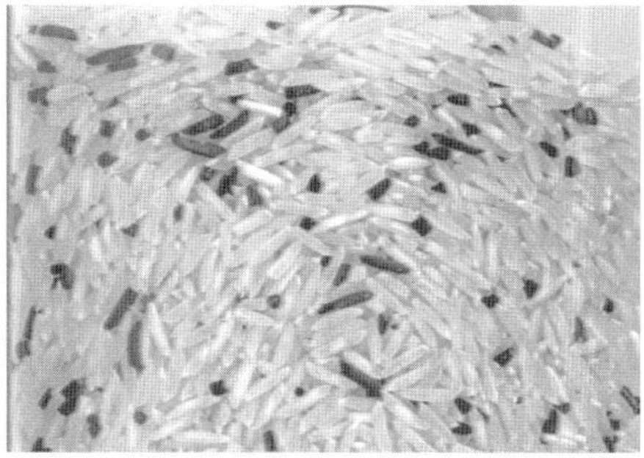

Figure 10. Snapshot of the rice packing during compaction. Structure of the convection roll can be seen at the wall

for spheres. We measured the rate of compaction and decompaction during each phases. We found for a tap intensity $\Gamma = 6$ a shock wave velocity $v = (8.7 \pm 0.5)$ m.s^{-1}. This value can be compared to numerical simulations carried out on quite a different system : spherical glass beads. In [18] , Gray and Rhodes observed such a shock wave and found a velocity of between 10 and 100 m.s^{-1}. The use of a high speed camera also allows us to extract more information such as the dependence of dilatation on the intensity of taps. This work is still in progress.

The behavior of our system is totally different for taps of intensity $\Gamma < 3$. After a transient the convection is extremely slowed down by compaction. For basmati rice and for a perturbation of intensity $\Gamma = 2.4$ the convection is totally stopped in most of the packing after $\tau \simeq 1500$. Convection rolls do not exist anymore, except near the air-media interface. Another interesting point is that an ordering process is observed at the bottom of the container : the convection rolls compact the lowest part of the packing and align the rice grains horizontally. This order creation is different from that observed by Villarruel et al. [17] : the mean main orientation of the grains is horizontal and not vertical. The evolution of packing ratio during this experiment is showed figure 12. On this figure, it can be seen that the fit from Chicago's law does not describe the experimental evolution very well. Indeed, the Chicago's law increases too fast for low number of taps and reaches the steady state to late. The parameters of the Chicago's law are $\tau \approx 375$ and $B \approx 2$. The parameter τ for $\Gamma = 2.4$ is smaller than τ for $\Gamma = 6$, even when the evolution slows down with the decrease of Γ. This characteristic time which comes from Chicago's fit isn't physical. This clearly shows that the Chicago's law, contrary to the KWW law, does not describe our results. Like for the high intensity perturbations, we find that the time extracted from KWW fit $\tau \approx 1380$ ($\beta \approx 0.31$) and the characteristic time of the convection rolls are similar. τ can be correlated with a real time of evolution. The resolution of the fast camera (256×240) is not high enough to measure the shock wave velocity for $\Gamma < 3$.

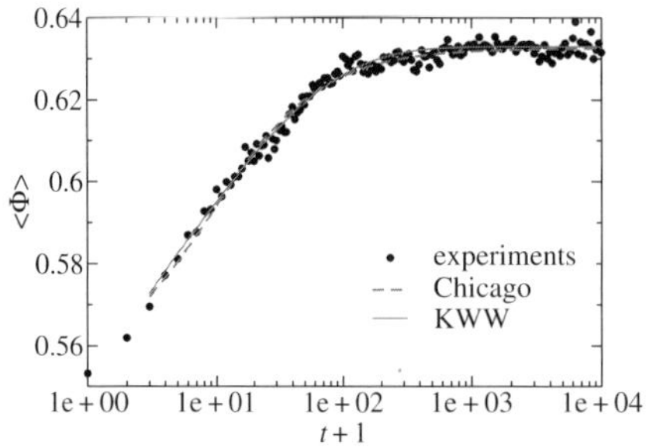

Figure 11. Temporal evolution of the mean volume fraction $\langle\Phi\rangle$ for $\Gamma = 6$. The experimental points are averaged on two experiments.

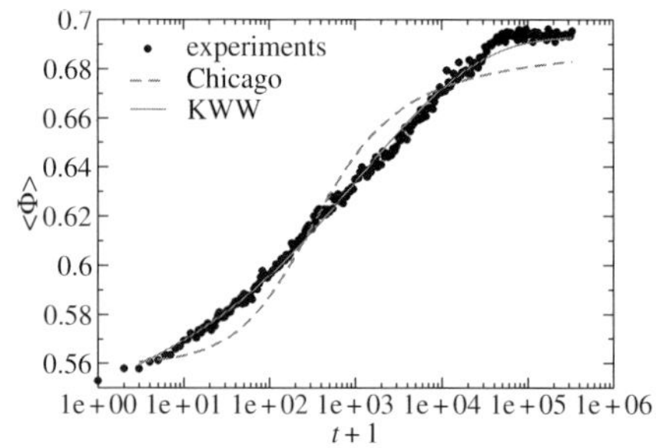

Figure 12. Temporal evolution of the mean packing fraction $\langle\Phi\rangle$ for $\Gamma = 2.4$. The experimental points are averaged on two experiments.

5. Discussion

In the previous section we presented experimental results on granular compaction under tapping. The next step of this study is to understand what happens on the grain scale. Indeed most previous experimental studies on granular compaction deal with the properties of, at least, a part of the packing (such as the mean packing fraction or its vertical profile). This gives only indirect information on grain behavior or on the packing microstructure. Experimental grain position determination during compaction is not an easy task since granular media is opaque. Nevertheless, very recently attempts were made to access to such data : x ray microtomography [19] and index matching liquid imaging [20]. In this section we present experimental studies carried out on x ray microtomography and discuss the different hypothesis on grain motion

5.1. Packing microstructure

X-ray microtomography provides measurements of the 3D structure of mesoscale materials. Applied to granular matter [21] it allows one to access to the position of the grain. We recently used this technique to the study of granular compaction. The x-ray microtomography set-up is provided by the ESRF (European Synchrotron Radiation Facilities) ID19 beamline. A monochromatic coherent beam is used to get sample radiography for 1200 angular sample positions ranging from 0 to 180 degrees. An x-ray energy of 51 keV was selected to ensure a high enough signal to noise ratio. The exposure time is 1 second by projection. The energy is selected using a classical double monocrystal device. A filtered backprojection algorithm [22,23] is used to compute the three-dimensional mapping of the linear absorption coefficient in the sample. The light detector used is based on the FRELON CCD (Fast REad out LOw Noise Charge Couple Device) camera developed by the ESRF Detector group. The CCD array is made of 1024 by 1024 elements and is 14 bits dynamic. A thin scintillation layer deposited on glass converts x-rays to visible light. Light optics magnify the image of the scintillator and project it onto the CCD.

With such a set-up we obtain a resolution of 9.81 μm by pixel. X-ray microtomography then allows one to have a fine and precise 3D reconstruction of the packing. We use an experimental set-up close to previous ones [1,10,16]: 200-400 μm diameter glass beads are poured up to about 80 mm height in a 8 mm inner-diameter glass cylinder. The whole system is vertically shaken by sinusoidal excitation at a frequency of 70 Hz. The applied acceleration is measured by an accelerometer and the intensity of the vibration is characterized by Γ, the maximal applied acceleration normalized by gravity ($g = 9.81$ m.s^{-2}). The experiments are carried out as follows: starting from an initial loose reproducible configuration ($\Phi \approx 0.57$), a packing is vibrated for a given number N of oscillations with a fixed acceleration Γ and then analyzed by x-ray microtomography. Each reconstructed packing contains about 15 000 grains. An example is reported in figure 13. Then, using an image processing software, the size and location of each grain can be recovered.
Due to the long time needed to perform a three dimensional mapping (about 1 hour) and due to the large number of perturbations needed, the study of grain positions during each step of the compaction is unfortunately impossible. Nevertheless such technique allows one to compare the packing microstructure at different stages of the compaction. figure 14

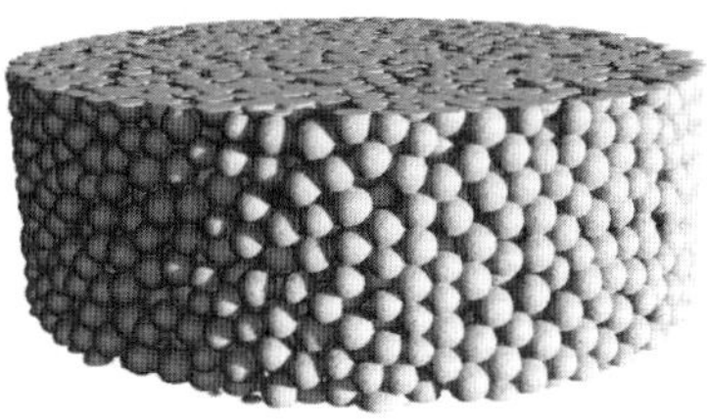

Figure 13. Part of a packing reconstruction obtained by x-ray tomography

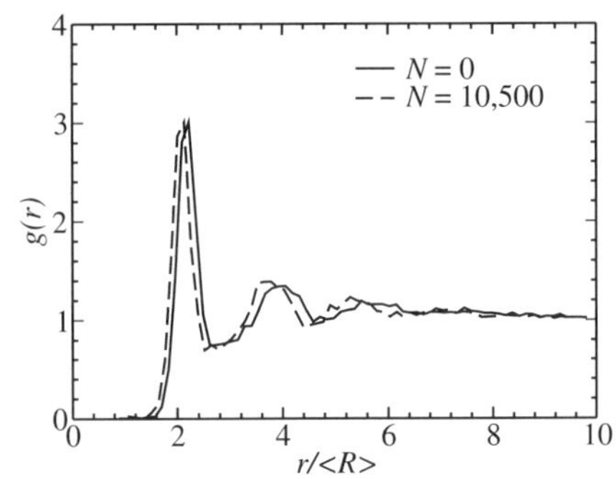

Figure 14. Evolution of the pair correlation function of $\Gamma = 3.0$

represents the variation of the well-known pair correlation function for the initial packing and after more than 10^4 excitations of intensity 3.0. Although the packing fraction increases during this compaction (from 0.57 to 0.62) no important signs of this compaction is seen on $g(r)$. This is even true for steady-state packing at different excitation intensity (see [19]). $g(r)$ is not a relevant parameter to study compaction. Further evidence of a transformation in the packing microstructure is provided by the size distribution of the interstitial voids. Moreover this can be an experimental test for the free volume theory that postulates an exponential decay for this size distribution $\rho(v) = \exp(-v/v_0)$. Following the work of Philippe and Bideau [9] we define a pore size via the Voronoï tessellation. A Voronoï polyhedron around a sphere is the region of space in which all points are closer to this sphere than to any other sphere in the packing. The Voronoï network, which is the whole collection of the edges and of the vertices of the polyhedra, maps the pore space. Each vertex of this tessellation is equidistant to four neighboring spheres and therefore defines the center of a pore. The volume of a pore is then the size of a virtual sphere centered on the vertex and in contact with the four neighboring spheres. The volume of this void-sphere partially represents the volume of the whole void volume situated inside the tetrahedron formed by the centers of four neighboring spheres. Since in our experiments the spheres are not perfectly mono-sized, we have adapted this method to a polydisperse packing replacing the Voronoï tessellation by the so-called navigation map [24]. It is then possible to compute the size distribution of the pores for each packing. Following Philippe and Bideau [9] the volume of the pores v is normalized by the mean volume of a grain $\langle w \rangle$. Since previous theoretical works on a free volume model ([8,3]) postulated an exponential decay for the distribution of the voids $\rho(v) \propto \exp(-v/v_0)$, we have reported our results in semi-log axis. Figure 15 presents the evolution of the volume distribution of the pores

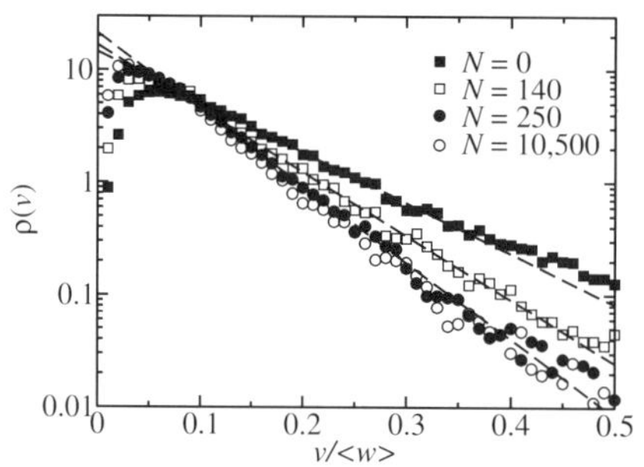

Figure 15. Evolution of the volume distribution of the pores during compaction with $\Gamma = 3.0$.

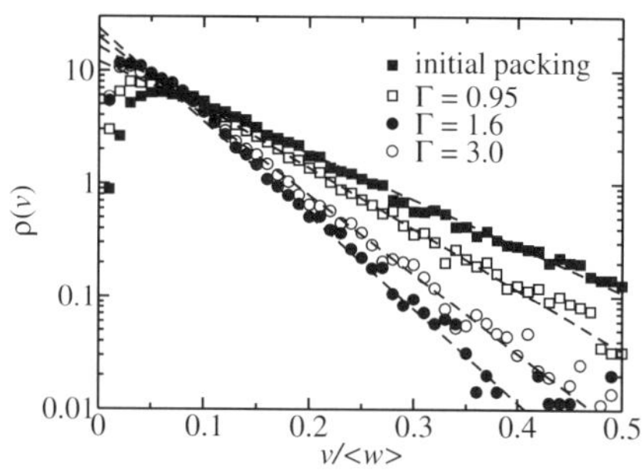

Figure 16. Volume distributions of the pores for the initial packing and for three different steady-state packings obtained for $\Gamma = 0.95$, 1.6 and 3.0.

for $\Gamma = 3.0$ at different stages of the compaction. First of all, we observe that unlike the pair correlation function, $\rho(v/\langle w \rangle)$ drastically changes with the number of excitations. Indeed an exponential decay law is found for the distribution of the voids, at least for the pores greater than octahedral pores ($v > (\sqrt{2}-1)^3 \approx 0.0711$). The exponential shape persists during the compaction of the packing, yet with a reduction of the tail, i.e. with a decrease of the characteristic volume v_0. As reported in figure 16, the final exponential decays of the steady-state distribution, obtained after a sufficient number of oscillations, are strongly dependent on the intensity Γ of the vibrations. By contrast, the part of the distribution corresponding to the smallest pores is less Γ-dependent. The remarkable point is that these results are very close to those obtained by numerical simulation by Philippe and Bideau [9] (see [19]). Their simulation is solely based on the geometric constraints in a three-dimensional packing of identical hard spheres. The agreement between this simulation and our experiments emphasizes the fundamental importance of the geometry in granular compaction. The steric constraint between the grains seems to rule the local rearrangements of grains allowed by the shaking energy.

5.2. Movement of the grains

As explained above, the use of x-ray microtomography is not actually an appropriate method to study the displacement of the grains. Very recently Pouliquen et al. [20] use an refractive index matching method to track the particles during compaction. Contrary to most granular compaction experiments where vertical taps are imposed to the granular packing these authors impose a periodic shear deformation to a granular packing in a box [25]. Another important difference between our set-up and Pouliquen et al's one is that the experiments are not carried out in air but in a liquid. The use of this fluid allows one to track particle motion during compaction. These authors found that the grain motion is not diffusive and exhibits a transient cage effect, similar to the one observed in glasses. In other words particle motion can be divided in two steps : a random motion within a cage whose extend is directly proportional to the shear amplitude and occasionally a change of cage. Pouliquen et al. also showed that contrary to cage motion, cage changes are irreversible phenomena and are directly linked to the compaction. An interesting work to carry out is to compare the compaction mechanisms in shearing experiments and tapping experiments. This can be done using numerical simulations described in [9,19] or experimentally using Pouliquen et al's method in our set-up.

6. Conclusion and open questions

In this article we present an extensive experimental and numerical study of granular compaction. Contrary to the Chicago group's well known experiment we use an experimental set-up with a large vessel. Indeed, the diameter of the cylinder is about 100 grain diameter as for Chicago group's experiment, it is about 10 grain diameters. We show that in this configuration, the granular compaction follows a KWW law and that this results does not depend on the grain anisotropy. We also find a KWW law for compaction of long shaped rice and short shaped rice. An important point is that the KWW law is also found in glass relaxation. These result are in disagreement with Chicago group's results. Is this due to the difference of vessel thickness ? With such a thin vessel Chicago group's experiments, contrary to ours, are carried out with a very limited convection. The use of x-ray microtomography allows the analysis of the reorganization of the grains and the distribution of the pore volumes. It was found in good agreement with numerical simulations of Philippe and Bideau [9] that compaction is mainly due to a decrease of the number of the largest pores. The volumes of these pores are statistically distributed along a broad exponential tail which progressively reduces while the structure gets more compact.

Acknowledgments

We thank Stéphane Bourlès and Patrick Chasles for technical assistance and Nicolas Taberlet for a critical reading of the manuscript. We are grateful to Xavier Thibault, Fabrice Barbe and Stéphane Bourlès for their contribution to the x-ray microtomography experiments. We acknowledge the European Synchrotron Radiation Facility (ESRF) for the use of their facilities, hospitality and financial help.

REFERENCES

1. J.B. Knight, C.G. Fandrich, C.N. Lau, H.M. Jaeger and S.R. Nagel, Phys. Rev. E, **51**, 3957 (1995); E.R Nowak, J.B. Knight, E. Ben-Naim, H.M. Jaeger and S.R. Nagel, Phys. Rev. E **57**, 1971 (1998).
2. E. Caglioti, V. Loreto, H.J. Herrmann and M. Nicodemi, Phys. Rev. Lett. **79**, 1575 (1997).
3. E. Caglioti, A. Coniglio, H.J. Herrmann and V. Loreto and M. Nicodemi, Physica A, **265**, 311 (1999).
4. M. Nicodemi and A. Coniglio, Phys. Rev. Lett. **82**, 916 (1999).
5. A. Barrat and V. Loreto, J. Phys. A, **33**, 4401 (2000).
6. J. Talbot, G. Tarjus and P. Viot Phys. Rev. E, **61**, 5429 (2000).
7. S.F. Edwards and D.V. Grinev, Phys. Rev E, 58, 4758 (1998).
8. T. Boutreux and P.G. de Gennes, Physica A, **244**, 59 (1997).
9. P. Philippe and D. Bideau, Phys. Rev. E, **63**, 051304 (2001).
10. P. Philippe and D. Bideau, Europhys. Letters, **60**, 677 (2002).
11. R. Kohlrausch, Pogg. Ann. Phys.Chem. **91**, 179 (1854)
12. G. Williams and D.C. Watts, Trans. Faraday Soc. 66, 80 (1970)
13. J.C. Phillips, Rep. Prog. Phys., **59**, 1133 (1996).
14. J.B. Knight, E.E. Ehrichs, V.Y. Kuperman, J.K. Flint, H.M. Jaeger and S. Nagel, Phys. Rev. E, 54, 5726 (1996)
15. P. Evesque and J. Rajchenbach, Phys Rev Lett. **62**, 44 (1989)
16. P. Philippe and D. Bideau, Phys. Rev. Lett. **91** 104302 (2003).
17. F. X. Villarruel and B. E. Lauderdale and D. M. Mueth and H. M. Jaegern Phys. Rev. E. **61**, 6914 (2000).
18. W. A. Gray and G. T. Rhodes, Powder Technology, **6**, 271 (1972).
19. P. Richard, P. Philippe, F. Barbe, S. Bourlès, X. Thibault and D. Bideau, Phys. Rev. E, **68**, 020301(R) (2003).
20. O. Pouliquen, M. Belzons and M. Nicolas, Phys. Rev. Lett., **91**, 014301 (2003).
21. G.T. Seidler, G. Martinez, L.H. Seeley, K.H. Kim, E.A. Behne, S. Zaranek, B.D. Chapman, S.M. Heald and D.L. Brewe, Phys. Rev. E, **62**(6), 8175 (2000).
22. G.T. Herman, Image Reconstruction from Projections. Academic Press, New York, 1980.
23. F. Natterer, The mathematics of computerized tomography. John Wiley sons, New York, 1986.
24. P. Richard, L. Oger, J.P. Troadec and A. Gervois, Eur. Phys. Jour. E, **6**, 295 (2001).
25. M. Nicolas, P. Duru and O. Pouliquen, Eur. Phys. J. E **3**, 309 (2000).

Unifying Concepts in Granular Media and Glasses
A. Coniglio, A. Fierro, H.J. Herrmann and M. Nicodemi (editors)

Experiments in randomly agitated granular assemblies close to the jamming transition

G. Caballero[ab], A. Lindner[a], G. Ovarlez[c], G. Reydellet[a], J. Lanuza[a] and E. Clement[a]

[a]Laboratoire des Milieux Désordonnés et Hétérogènes, Case 86, 4, place Jussieu, 75252 Paris Cedex 05, France

[b]Departamento de Física, Facultad de Ciencias, Universidad Nacional Autónoma de México, 04510 México, Distrito Federal, México.

[c]Laboratoire des Matériaux et Structures du Génie Civile (LCPC-ENPC), Cité Descartes, 2, allée Kepler, 77420 Champs sur Marne, France

We present here the preliminary results obtained for two experiments on randomly agitated granular assemblies using a novel way of shaking. First we discuss the transport properties of a 2D model system undergoing classical shaking that show the importance of large scale dynamics for this type of agitation and offer a local view of the microscopic motions of a grain. We then develop a new way of vibrating the system allowing for random accelerations smaller than gravity. Using this method we study the evolution of the free surface as well as results from a light scattering method for a 3D model system. The final aim of these experiments is to investigate the ideas of effective temperature on the one hand as a function of inherent states and on the other hand using fluctuation dissipation relations.

1. Introduction

Strikingly, systems as different as dense emulsions, colloidal pastes, foams or granular matter have many rheological properties in common [1]. All these systems can flow like fluids when a sufficiently high external stress is applied but jam into an amorphous rigid state below a critical yield stress. This jamming transition is associated with a slowdown of the dynamics which led Liu et al. [1] to propose an analogy between the process of jamming and the glass transition for glass-forming liquids. Although the nature of this jamming transition is still unclear experimentally [2], several attempts were made to adapt the concepts of equilibrium thermodynamics to athermal systems out of equilibrium [3–8]. For packings made of grains with a size larger than a few microns, thermal fluctuations are too small to allow a free exploration of the phase space. The grains are trapped into metastable configurations. The system can not evolve until external mechanical perturbations like vibration [9] or shear [10] are applied allowing the grains to overcome energy barriers and triggering structural rearrangements. In this case, the free volume and the configurations accessible for each grain are capital notions that were used to define the new concept of "effective temperature" [3]. It was proposed recently that this notion

could account for the transport properties in the vicinity of a jammed state through a fluctuation dissipation theorem [4].
A. Fierro *et al.* [11] treat this question in analogy with supercooled liquids. In this case we understand by inherent states those that do not evolve with time and that correspond to the local minima of the potential energy in the 3N-dimensional configuration space of the particle coordinates. A. Fierro et al. [11] now consider the mechanically stable states of granular materials at rest as inherent states. They work numerically with a 3D system of hard spheres subject to gravity and undergoing a Monte-Carlo shaking. During the dynamics, the system cyclically evolves for a time τ_0 (corresponding to the tap duration) at a finite value of the bath temperature T_Γ (corresponding to the tap amplitude) and is suddenly frozen at zero temperature in one of its inherent states. They find that the system reaches a stationary state determined by the tap dynamics (i.e. different T_Γ and τ_0). These stationary states are indeed characterized by a *single* thermodynamical parameter since one finds a single master function when the fluctuation $\overline{\Delta E^2}$ is plotted as a function of $\overline{E}$, where E is the potential energy of the ensemble of grains. Based on this result, they conclude that the quasi-stationary state can be genuinely considered a "thermodynamical state" and they define an effective temperature through the fluctuation-dissipation relation.
Here we design an experimental set-up with the final aim to test closely the ideas of effective temperature on the one hand as a statistics of inherent states and on the other hand as a result of the fluctuation-dissipation relation by studying the transport properties of a grain. Contrarily to previous work on vibrated granular assemblies, we design a new way of shaking the granular material at accelerations much lower than gravity. In the first part of the paper, we study a model granular assembly in 2D and use this preliminary investigation to design the final 3D experiment. The preliminary results obtained with this set-up are presented in the subsequent section.

2. Vibration of a 2D model granular assembly

2.1. Experimental Set-Up

We study the displacement of tracer particles in a 2D model granular assembly which is exposed to tapping. To do so we use the following model system (figure 1): a layer of polydisperse particles is confined between two vertical glass plates. The dimensions of the glass plates are 20 cm times 30 cm. The lateral walls are made of Teflon. We use a mixture of small cylindrical steel particles of 3 mm height with three different diameters, notably $d_1 = 6$ mm, $d_2 = 5$ mm and $d_3 = 4$ mm to create a disordered packing. The cell is partially filled which leads to about 30 particles in the vertical direction and to 60 particles in the horizontal direction.

We now study the response to tapping of the system by using an electromagnetic shaker. A single sinusoidal tap is applied every 5s. The subsequent motion of the shaker and the beads lasts about 1 s. Note that we can vary the intensity of the tapping by changing the applied voltage from 100 mV to 500 mV leading to accelerations from approximately $\gamma_{peak}/g = 1$ to $\gamma_{peak}/g = 2$, with γ_{peak} being the peak acceleration. The system contains between 1 to 3 tracer particles of diameter $d_1 = 5$ mm that can be positioned at a given initial position. A CCD camera coupled to a computer captures a picture after a given

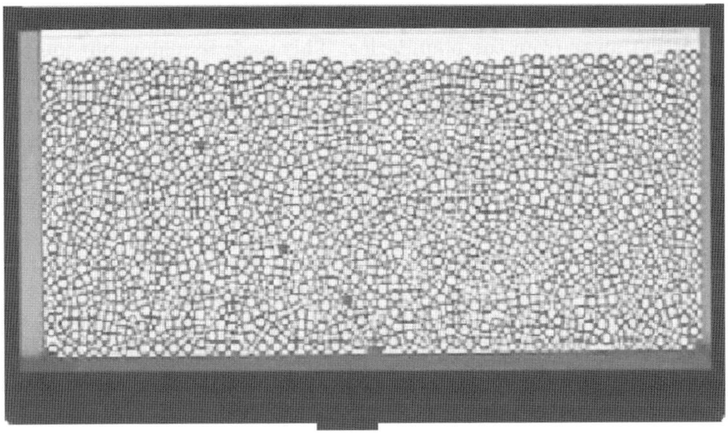

Figure 1. Experimental set-up showing the 2D model system for a granular assembly under vibration: a layer of polydisperse particles is confined between two vertical glass plates. A single sinusoidal tap is applied every 5 seconds via an electromagnetic shaker.

number of taps (typically between 500 to 1000 taps) and allows us to follow the long time displacements of the tracers. We then extract the trajectories of the tracer particles using image processing and finally obtain their x and y coordinates as a function of time.

Note that we typically observe a compaction of the initial surface occupied by the grains of about 1% during the first 10000 taps, that correspond to the very beginning of our experiments. Afterwards we observe fluctuations of the height of the granular layer but can not detect further compaction within our experimental resolution.

2.2. The large scale and long time convection dynamics

In the following, we describe the experimental results obtained when performing experiments with a free surface at the top of the granular assembly. In this case, for accelerations larger than the acceleration of gravity and for a given phase of the motion, the grains on the upper surface are launched freely with an upwards velocity given by the acceleration of the cell. A bit later an impact with the rest of the granular assembly occurs as the latter catches up with the grains at the upper surface [12]. In the absence of boundaries all grain would follow about the same free flight trajectories. This holds only if the grains come to a complete rest between two taps and thus provided that the time for energy dissipation is smaller than the time between two impacts. The sequence of impacts is at the origin of the more or less random shaking of the granular assembly and subsequently, is triggering the compaction/decompaction phenomenology.

The problem is that this shaking procedure is strongly complicated by the presence of frictional boundaries which are in our case the lateral walls made of Teflon. The motion of the grains in contact with the boundaries is perturbed by friction and for vertical boundaries these grains hit the rest of the assembly slightly before their neighbors positioned further away from the boundaries. This leads to a slow but inexorable descent of the grains at the boundaries since these grains are likely to occupy the empty space left by their neighbors still in free flight. Because of mass conservation, the compound of this motion, tap after tap, leads to large scale convection rolls [13–15].

Furthermore, the magnitude of the convection rolls depends not only on grain/boundary friction values [15] but also on the packing density values. The reason is that the higher the packing fraction, the more efficient is the transmission of vertical forces to horizontal forces (given by the Janssen's effective parameter : see [16]). This effect, in association with the friction coefficient, fixes a limit between an upper part where the convection rolls are located and a bottom part which is blocked since the boundary grains cannot overcome the Coulomb threshold. It was show by a simple argument that this effect of a jammed phase localization depends on the aspect ratio of the cell, it is reduced in the limit of small friction and is bound to disappear for maximal accelerations larger than $2g$ [17].

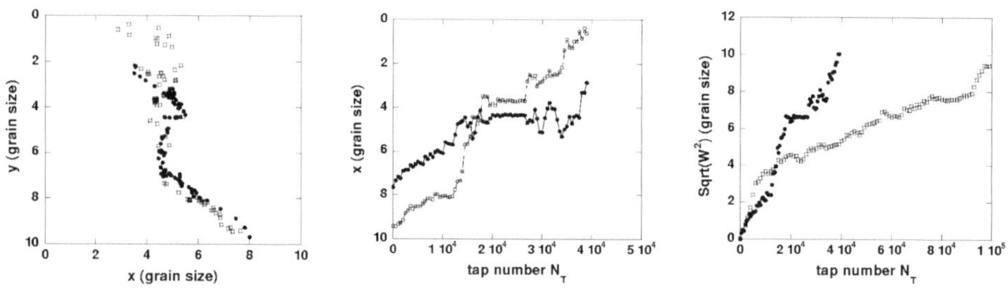

Figure 2. Tracer displacement for two experiments having the same initial configuration but different tapping amplitudes, given by the applied voltage of 150 mV for experiment 1 and 325 mV for experiment 2. For experiment 1 a snapshot is taken every 1000 taps and for experiment 2 every 500 taps. All displacements are scaled on a typical grain size of 5 mm. The observed tracer is placed in the middle of the celle at the beginning of the experiments. Left: Trajectory of experiment 1 (□) and experiment 2 (•). Middle: x (•) and y (∘) coordinate for experiment 2 as a function of the tap number. Right: $\sqrt{\langle W^2 \rangle}$ as a function of tap number for experiment 1 (□) and experiment 2 (•)

First, we describe two experiments having the same initial configuration but different tapping amplitudes. The steel grains are chosen to have a low friction with the Teflon boundaries such that for all the experiments we show, we do not observe a phase where the convective motion is blocked in the bottom part. The experiment with the lower tapping amplitude is experiment 1 whereas the experiment at a higher tapping amplitude is experiment 2. The left graph of figure 2 shows the trajectories of the tracer for the two experiments. The displacement is scaled on an average grain size of 5 mm. One observes that the tracers follow in both cases the same trajectory, signature of a convective motion without significant diffusion of the particles. Note that for experiment 1 a snapshot (corresponding to one data point) is taken every 1000 taps whereas for experiment 2 a snapshot is taken every 500 taps. The velocity of the tracer is thus different from one experiment to another as one would expect. Even if the two trajectories seem quite smooth, the displacement of the tracer as a function of time is very irregular. This is illustrated on the graph in the middle of figure 2 that shows the x and y coordinates of experiment 2. We have observed for several experiments, that the tracer does nearly not move for a large number of taps and then abruptly continues its displacement. The graph on the right side of figure 2 finally shows $\sqrt{\langle W^2 \rangle}$ the sliding average of the root mean square displacement for experiments 1 and 2. This graph shows clearly that the displacement as a function of the number of taps is less important for experiment 1 than for experiment 2. Furthermore, it becomes again clear from this graph that large scale convective displacements are taking place in the granular assembly. An estimation of the average particle velocity observed in our experiments leads to values roughly between $v_{particle} \sim 1$ particle size per 1000 taps or even per 10.000 taps depending on the applied voltage. Interestingly, for grains of typically the same size, Philippe et al. [18] find the same order of magnitude for their 3D tapping experiment. Note, that the convection velocity increases with the tapping amplitude.

The importance of the large scale displacements becomes even more clear when looking at a third experiment (experiment 3). In this case, the tapping amplitude was increased further and three tracer particles were followed. A snapshot was taken every 500 taps. On the left of figure 3 one can see the trajectories of two of the three particles and one can conclude that huge convection rolls form in the system. Note that the convection roll observed occupies half the size of the cell. After one period the particle follows nearly the same trajectory again. This is illustrated on the right graph of figure 3 that shows the x and y coordinates for one of the particles as a function of the number of taps. One may notice that they follow nearly a perfect sinusoidal motion over more than one period. Once again one can conclude that the motion is purely convective.

Consequently, it is clear from these model experiments that the leading dynamical behavior for the shaken grains is convection with grains following well defined trajectories. Even if for low shaking amplitudes we have intermittent dynamics as seen on figure 2 (middle) we did not observe any significant diffusive motion of the grains. In other words, the grains are likely to keep their neighbors for a very long time. Thus, any attempt to characterize the compaction dynamics by self diffusive properties of the grains, is doomed to fail. Another important question that one could rise, is whether the observed convection dynamics is related to the steady states reported experimentally in tapping experiments [9,18–21]. One might suggest that the convection rolls lead to a decompaction of the system and that there is thus a competition between this decompaction and compaction due to shaking. This could explain the different results reported by Nowak and Knight *et al.* [9,20,21] and Philippe et al. [18,19] since they use columns with different aspect ratios. Moreover, the column by the Chicago group is very narrow (about 15 grains across) and the role of wall friction and finite size effects could then be crucially important. All these questions are to us still open.

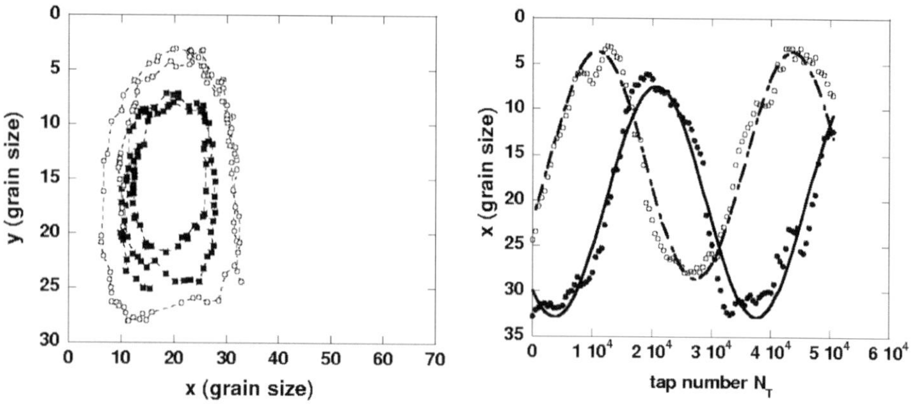

Figure 3. Tracer displacement for experiment 3 with a high tapping amplitude (500 mV). A snapshot is taken every 500 taps. Left: Trajectories of tracer 1 (■) and tracer 2 (○) showing huge convection rolls. Right: x and y coordinate of tracer 2 as a function of N_T.

2.3. Suppression of convection

We now try to suppress the convection by putting a lid on top of the cell at a distance of less than one bead diameter. The idea is to produce a bounce of the grains on the lid almost immediately after the launch of the packing. This bounce downwards is likely to suppress the time lag effect in the grain trajectories due to the presence of walls with friction.

The results obtained are indeed significantly different from those with a free surface. A direct visualization of the packing witnesses a rather strong agitation of the grains: one can note rather pronounced collective motions of the grains where large assemblies of particles seem to oscillate very slowly. A closer look at the trajectories (figure 4 left) shows however that these trajectories are clearly localized. Note that all displacements shown on this graph are less than one particle diameter, even for very long times. Note that the bottom tracer is close to the bottom plate which explains its slightly stronger agitation. Figure 4 (right) shows the trajectory of the upper tracer in detail and one can conclude once more that the displacement of the grains are very small. From figure 5 that shows $\sqrt{\langle W^2 \rangle}$ one can conclude that once again we do not observed any significant self diffusion of the particles but that in this case the displacement is localized. Note that the strongest displacement is seen for the bottom tracer as explained above.

Therefore, when the convection is suppressed, we still observe modes of motion of the packing that can lead to reorganization and compaction but it is clear now that those modes are collective and are not associated to self-diffusion properties of a grain.

2.4. Outlook

In the previous sections we have shown that when applying strong tapping ($\gamma_{peak}/g > 1$) to a 2D granular assembly, strong convection takes place. This convection can be

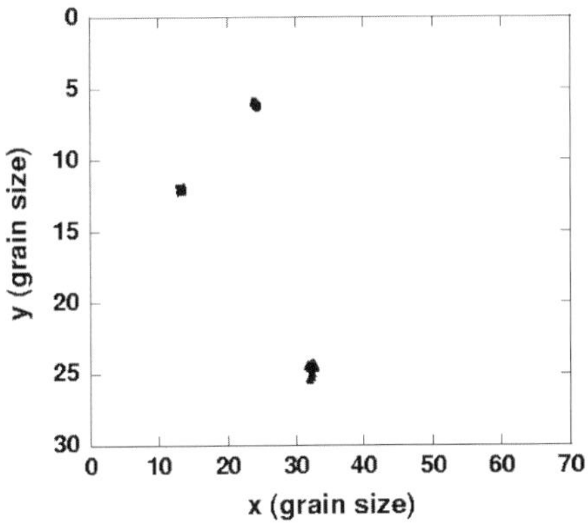

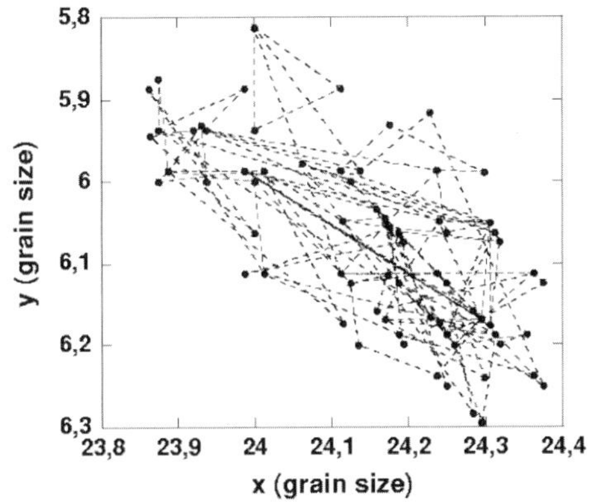

Figure 4. Displacement of three tracer particles for experiment 4, having the same initial configuration and tapping amplitude as experiment 3, but with a lid on top. Left: Trajectory of three tracer particles. Right: Trajectory of one of the tracer particles (the upper particle) in detail.

suppressed by putting a lid on top of the granular assembly, but then the position of the lid has to be finely tuned. A lid too close to the packing may suppress drastically the granular motion and for a lid too far away may start convection again. Clearly, this is not a method suited to study a problem where compaction i.e. the packing height, may vary.

Thus, we suggest to study a different system, that might allow to suppress convection in a more adequate way and might thus be better adapted to study the dynamics and the transport properties of an agitated packing of grains. To do so, we have built a set-up where a large number of individually controlled pistons are moving up and down at the bottom of a cell identical to the one described in section 2.1. The individual control of the pistons allows to go from spatially and temporally coherent to more complex agitations. Furthermore, lower tapping amplitudes can be applied such that $\gamma_{peak}/g < 1$ which will also suppress convection. We plan to measure the displacement of one or several tracers as well as the collective modes of reorganization, and to measure the mobility of a tracer when a given constant force is applied. A way to do this is to use a denser particle or design a way to pull the tracer. Further more we plan to measure the fluctuations of the free surface. The results will then be compared to the results obtained in 3D that will be described in the following.

3. Compacting a 3D granular assembly

The experimental set-up in 3D is set to create in the bulk a random agitation of the grains at low level of energy and in conditions where convection rolls due the uplift of the grains are completely suppressed. Following the results of the 2D model system

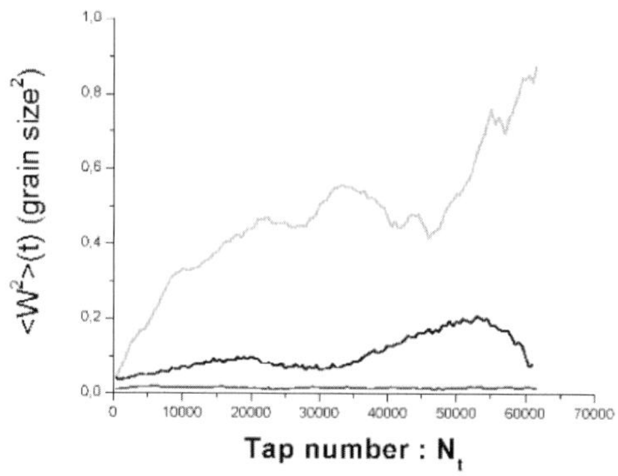

Figure 5. $\sqrt{\langle W^2 \rangle}$ as a function of tap number for the three tracers of experiment 4. The two curves with the lower displacement correspond to the two upper tracers whereas the curve with the highest displacement corresponds to the bottom tracer.

presented in the previous section, we seek to produce an influx of energy creating a quasi randomly agitated surface in connection with the bulk of the packing. This is the closest we could think of a "thermal bath". The practical method is described in the next subsection. Then, we expose some preliminary results showing that the method is encouraging and might well be suited to study compaction and jamming processes in granular assemblies and also leads to a path where the notion of effective temperature can be precisely addressed.

3.1. Experimental set-up

The vibration device is made of five piezoelectric transducers commercially sold as a part of a tweeter. As shown in figure 6, the transducers are at the bottom of a conical shape paper container which is filled with $1.5mm$ glass beads. The surface of the granular packing confined inside the tweeters is fixed to the observation cell and constitutes its bottom. The cell is a hollow rectangular box of $20cm$ length, $2.3cm$ width and around $4cm$ high. The front and rear boundaries are made of glass. The piezos are excited by a $380Hz$ square signal with a maximum effective voltage $V_{eff} = 35V$ (figure 6). The resonance frequency of each piezo is $f_0 = 1200HZ$. Note that the power that each piezo dissipates is enough to make a single and lonely grain on the piezo fly up to $5mm$ high. However, when the box is filled with grains (around $350g$), the grain motion induced by the piezos is so small that it is not perceptible to the eye.

3.2. Measurement techniques

The cell is instrumented to measure several things (see figure 7) : (i) an inductive probe provides the position of a thin metallic plate lying on the top granular surface and gives

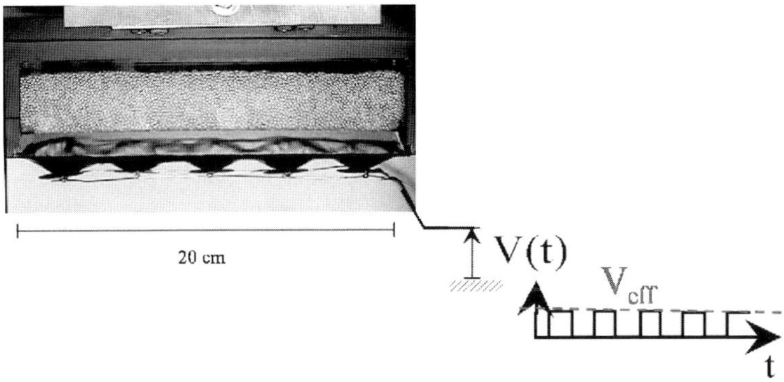

Figure 6. Experimental set up: five piezoelectric transducers constitute the bottom of a rectangular glass box which is filled with $1.5mm$ glass beads. The piezos are excited by a $380Hz$ square signal with a maximum effective voltage $V_{eff} = 35V$

information about the mean **compaction or the potential energy** of the packing, (ii) an accelerometer buried in the granular packing provides informations on the **effective injected power** (iii) we use a Diffusing Wave Spectroscopy (DWS) technique to measure the **characteristic times involved in the microscopic dynamics** [22], (iv) finally, the cell is also designed to evaluate the **mobility** of an intruding grain. This last point will be described more precisely in the last subchapter.

The basic idea of DWS is to make a coherent light (laser) pass through a scattering sample in which each photon is elastically scattered multiple times in such a way that it performs a random walk. Under this circumstances, the light diffuses through the sample and forms an interference pattern after it. The microscopic dynamics of the grains are related to the dynamics of the light interference pattern. Here, a multispeckle technique is used that was developed earlier for colloids. The system dynamics assessment is made through the computation of the light intensity autocorrelation function:

$$g_2(t_w, t_w + t) \equiv \frac{\langle I(X,Y,t_w)I(X,Y,t_w+t)\rangle_{X,Y}}{\langle I(t_w)\rangle_{X,Y}\langle I(t_w+t)\rangle_{X,Y}} - 1 \qquad (1)$$

where t_w is the time of the reference image, t is the time relative to t_w and (X, Y) are the spatial coordinates on the speckle image. The theory of light diffusion shows that under the regime of multiple scattering (where the photons perform a random walk) it is possible to quantitatively relate the characteristic correlation time of g_2, τ, to the mean displacement of the grains (scatterers) [22]. However, in our case, preliminary measurements show that due to the relatively small number of grains across the cell (about 14 grains), the condition of a fully developed random walk for the light path is probably not fully achieved and thus, the possibility to trace back quantitatively the average motion of the grains is somehow complicated. On the other hand, at this stage,

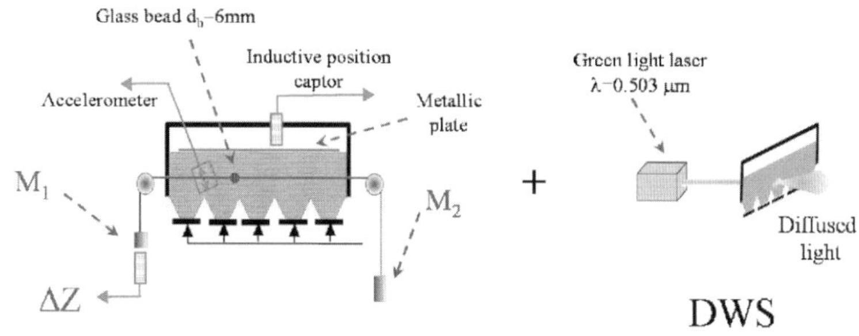

Figure 7. We can measure in the system: the compaction or potential energy, the effective injected power, the characteristic microscopic time (DWS technique) and the mobility of a grain.

the correlation function is still a good and sensitive indicator for the granular dynamics.
To prepare the system, we poured the grains into the box and gave gentle taps to compact the grains such as to "erase" the structural memory of the pouring and to prepare the system at a given packing fraction. Next, we placed the thin metal lid on the surface and let the laser go through the packing. We left it go for one hour in order to stabilize the laser as well as the initial stress relaxations in the sample. Note that in these experiments we did not put the intruding grain because it would have interfered with the other measurements. Once stabilized, we turned on the piezos at their maximum capacity ($35V_{eff}$, which corresponds to a r.m.s. acceleration measured by the accelerometer of $\langle\gamma^2\rangle^{1/2} = 0.24m/s^2 << g$). After an hour of vibration we changed the voltage value. We continued the vibration changing the voltage each hour, measuring the position of the surface and recording the speckle images during each of these one-hour vibration intervals. Actually, we made two different experiments (labeled experiment 1 and experiment 2) with different voltage variation sequences.

3.3. Preliminary results

3.3.1. Experiment 1 - compaction and microscopic dynamics

The inner frame of figure 8 (left) shows the voltage variations that we followed in this experiment. We began with a *first descent* from $V_{eff} = 35V$ to $V_{eff} = 27.2V$; then, we came back to $V_{eff} = 35V$ and made a *second descent* down to $V_{eff} = 10V$. We did this with the idea of reaching a stationary state during the first strong vibration steps and to compare the measurements for the same voltage values in the second voltage descent. The correlation functions $g_2(t_w, t)$ of figure 8 (left) fit well with a stretched exponential $y(t) = exp(-(t/\tau)^\alpha)$.
On the right of figure 8 we can see the behavior of α and τ as a function of voltage. From the comparison of the first and second descents we don't observe any ageing effect, since the data corresponding to both of them are similar. Noteworthy, for the value of α

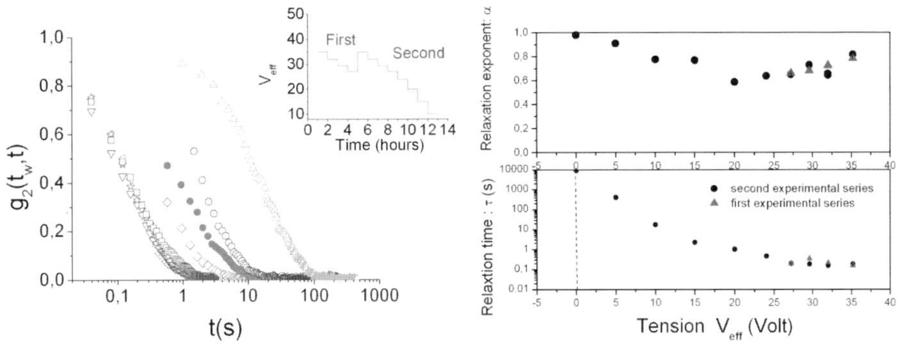

Figure 8. Left: Correlation functions for the second descent of experiment 1: $(\triangleleft)V_{eff} = 35V$, $(\triangledown)V_{eff} = 32V$, ($\bigstar$) $V_{eff} = 29.6V$, $(\square)V_{eff} = 27.2V$, $(\Diamond)V_{eff} = 24V$, $(\bullet)V_{eff} = 20V$, $(\circ)V_{eff} = 15V$, $(\triangle)V_{eff} = 10V$. Inner frame: Voltage variations in experiment 1: we began with a first descent from 35 to $27.2V$; then, we came back to $V_{eff} = 35V$ to make a second descent down to $V_{eff} = 10V$. Right: Behavior of α and τ, parameters of the fitting stretched exponential to the correlation function $g_2(t, t_w)$, as a function of effective voltage for the first and second voltage descents of experiment 1.

other authors [23] have reported a constant value of 0.8 for similar systems.

From the position of the surface we can, in principle, calculate the packing fraction ν and the potential energy E of the system. However, despite the high precision data that we have for the surface position, we can only roughly estimate the corresponding *absolute* value of the packing fraction (5% error). This is due to the lack of precision in determining the total volume of the cell+tweeters. The left of figure 9 (a) shows the compaction dynamics of the system during the first hour of vibration. After an initially "fast" decrease of the height, we indeed observe a slow logarithmic-like compaction process. Note that a stationary state is not yet reached. Plot (b) of the left of figure 9 shows a typical measure of the position of the surface during vibration at $V_{eff} = 35V$. It corresponds to the second descent, i.e. five hours after the beginning of the vibration and though it seems that the system had reached a stationary state, there is still a slow systematic compaction not perceptible at such time scale. Interestingly, the position of the surface seems to undergo wild fluctuations expanding and compacting in an irregular manner. However, the spatial displacement of the level of the surface is of the order of microns, much less than a grain size. So here we are talking about a collective macroscopic behavior since individual grain migration is not possible. This surface dynamics should be associated with the typical size of grain rearrangements in its surrounding (slipping of contacts and rotations) performed more or less randomly per unit time and integrated over the cell height. This is why we now investigate the dynamics of the fluctuations more precisely.

From the signal of figure 9 (b) we calculated the roughness $W = \langle (z - \langle z \rangle)^2 \rangle^{(1/2)}$, where z is surface's position, as a function of time and we found that the surface fluctuates as

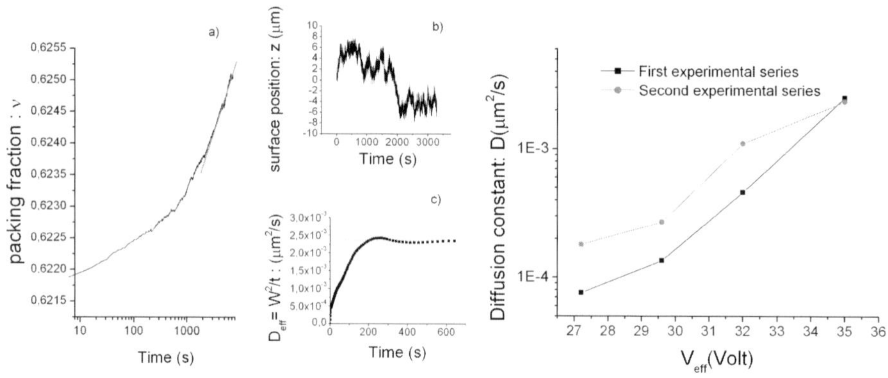

Figure 9. Left: a) Initial compaction at $V_{eff} = 35V$. A steady state is not observed. b) Short time diffusion-like fluctuations of surface's position during vibration. It corresponds to the second descent at $V_{eff} = 35V$. c) Surface's position fluctuates following a simple diffusion; the diffusion constant is calculated from position's roughness. Right: Effective diffusion of the height of the surface vs effective voltage

a simple diffusion process since $D_{eff} = W^2/t$ tends to a constant, the effective diffusion constant, for long times (figure 9, c)). In figure 9 right we show this constant for different values of the effective voltage for both experimental series. Now it is clear that there is a difference between the first and the second descent which could be a signature of the ageing of the packing. However a final conclusion is not possible here since the data corresponds to a single realization.

3.3.2. Experiment 2 - Dynamical intermittency

One of the first things we noticed when we began to work with vibrated systems was that the speckle dynamics during vibration was not continously varying but would rather show intermittent or irregular behavior. For a given vibration intensity we observed that the speckle changed with a typical characteristic decorrelation time, but the value of this time would often present big variations depending on which initial speckle pattern was chosen. Basically there were phases where the decorrelation would be regular and steady but others where an intense dynamics would drastically accelerate the speckle decorrelation. Interestingly, this kind of behavior was observed recently in thermal systems near jamming transition by Cipelletti *et al.* [24]. Moreover, these authors have proposed a way to quantitatively characterize the intermittency using the autocorrelation intensity function (eq. 1) obtained from the DWS technique. The idea is to keep fixed the time interval between correlating images t and to vary the reference image t_w. If the speckle dynamics were continuous this analysis would result in a kind of noise around a constant correlation value depending on t. However, huge "catastrophic" events would give low correlation values, highly deviated from the mean. The statistical analysis of such events leads to a quantitative characterization of the systems dynamics [24]. Following these ideas we

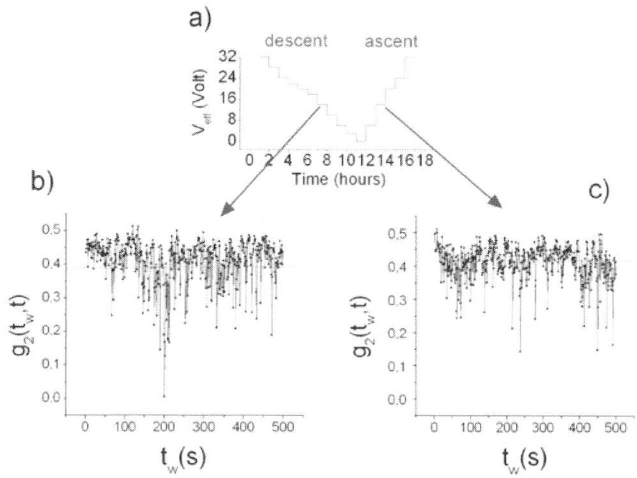

Figure 10. a) Voltage variations in experiment 2. b) Autocorrelation intensity function (eq. 1) for variable t_w and constant $t = 0.5s$ at $V_{eff} = 14V$ of the descending part of voltage variation. c) The same as (b) but for the ascending part. The correlation drops off from its mean value as a result of the intermittent dynamic of the system near jamming transition.

made an experiment similar to the previous one but with different voltage variations (fig. 10 (a)). In (b) and (c) of figure 10 we observe the autocorrelation intensity function for variable t_w and constant $t = 0.5s$ for the same vibration intensity ($V_{eff} = 14V$), (b) corresponding to the descending part of the voltage variation and (c) to the ascending part. It is possible to appreciate how the correlation frequently drops off from its mean value. In figure 11 is shown the histogram of the data from plots of figure 10. If there was an ageing effect the two histograms should be different, which is not the case. However, the deviation of the histograms from a Gaussian one to the low values of $g_2(t, t_w)$ is an evidence of intermittency [24].

3.3.3. Mobility of an intruding grain

In this experiment we tried to measure the mobility of a grain inside the vibrated granular system by forcing an intruding grain to move trough it. We glued the intruding grain to a fishing thread which passes through the granular assembly and that is tightened by two masses of different weight in an Atwood's machine-like configuration (see figure 7). The intruder's position is monitored by an inductive captor that measures the displacement of one of the tightening masses. The size of the intruding grain is of $6mm$ in diameter. Figure 12 shows three different regimes of the mobility as a function of $\Delta M = M_2 - M_1$

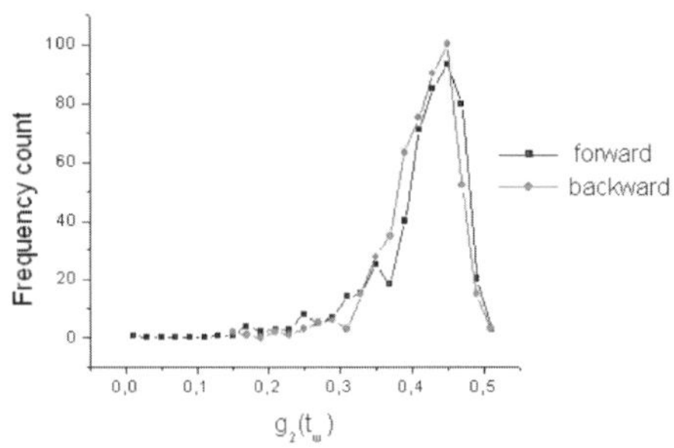

Figure 11. Histogram of the correlation curves of figure 10. The deviation from a Gaussian curve results from intermittency [24]. From comparison of the forward and backward data we don't find evidence of ageing for this particular case.

(see figure 7). For $\Delta M = 9g$ the intruding grain is initially moving but suddenly gets jammed. A continuous movement around a constant velocity is observed for $\Delta M = 15g$. Finally, for ΔM around $25g$ there is a plastic yield and the intruding grain moves without resistance through the granular material. Although these are only preliminary results they show that this technique allows for interesting mobility measurements.

4. Conclusions

We have shown the results of two experiments of vibration of a granular material, one in 2D and one in 3D, where both macroscopic and microscopic features of the flow can be measured and eventually related. In the case of the two dimensional assembly with free surface, it was also clear that the long time dynamics is dominated by convection and that dispersion due to local self diffusion is not a dominant feature. While moving into the huge convection rolls, the grains would roughly conserve their neighbors for a very long time. We could not measure any self-diffusion characteristics for the whole durations of the experiments and this conclusion still holds even when convection (but not the granular agitation) is suppressed by imposing a lid on the top. Consequently, from these direct observations in 2D, we came to the conclusion that in all standard shaking experiments, the observed compaction steady states could result from a competition between compaction due to vibration and expansion due to convection. Moreover, we stress on two important points. First, the suppression of convection is quite important to a controlled study of compaction dynamics and to access the notion of "effective temperature"more easily. Second, in the very dense phase, the notion of self-diffusion of a particle is unlikely to be a relevant parameter suited to characterize the jammed dynamics. Basically, the motions of the grains are localized in the cage formed by their neighbors and their jumps statistics within the cage reflect *collective* modes of reorganization. These dynamical processes are

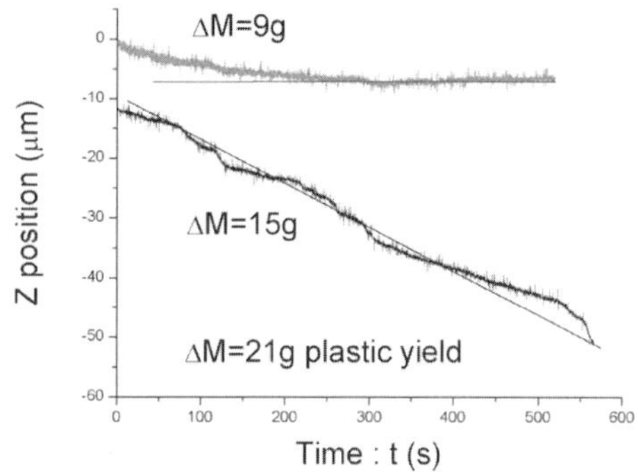

Figure 12. Measurement of the displacement of the intruding grain during vibration for different ΔM.

indeed sufficient to create global compaction/decompaction processes without motion of a grain out of its neighbor's cage.

We managed to achieve a convection free vibration in the three dimensional experiment by putting the grains in direct contact with five piezoelectric transducers. Then, we were able to study the compaction dynamics in the weak vibration limit, i.e. imposed by accelerations much smaller than gravity. A study of compaction under this circumstances is quite new. We observed a very slow density relaxation and diffusion of the free surface that we will be able to characterize more properly in experiments actually in progress. Interestingly, the magnitude of volume or density fluctuations indicates that there is no grain diffusion during vibration, thus the compaction and reorganization dynamics of the bulk is solely due to collective features. Another advantage of weak vibrations is that we managed to observe and measure intermittent dynamics which creates a link with results on jammed colloidal phases. This becomes important since, as pointed out by L. Cipelletti et al. [24], temporal heterogeneities are a fundamental feature of the dynamics of jammed systems. Interestingly, surface diffusion as well as intermittent dynamics were also found in stochastic compaction models [25,26]. Moreover, it has been proposed that a practical way to experimentally measure effective temperature is to analyze intermittent events [27]. These preliminary results are encouraging. We think that we will be able to extract very interesting information from this system once we had done all the calibrations needed and improved our procedures. Also, a central point is to manage to reach the stationary state, which seems not easy since we work with extremely weak vibrations. In fact, one of the things that we want to do is to increase the power injection to the system and to randomize it as much as possible. Certainly, it would be very interesting to couple together all the different measures that we can make on the system. With this, we could try to relate the microscopic features (DWS and mobility measurements) to macroscopic ones (compaction dynamics and density fluctuations). Of course, the next obvious step is to obtain the measurements of density and its fluctuations (or, equivalently, energy and its fluctuations) to compare them to the numerical results of A. Fierro et al. [11].

Acknowledgments This project is part of ECOS M03P01 and GC is supported by CONACYT and DGEP.

REFERENCES

1. A. J. Liu and S. R. Nagel, Nature **396**, 21 (1998).
2. F. da Cruz, F. Chevoir, D. Bonn and P. Coussot, Phys. Rev. E **66**, 051305 (2002).
3. S. F. Edwards, Physica A **249**, 226-231 (1998)
4. A. Barrat, J. Kurchan, V. Loreto and M. Sellitto, Phys. Rev. Lett. **85**, 5034 (2000); Phys. Rev. E, **63**, 051301 (2001).
5. F.K. Ono, C.S. O'Hern, D.J. Durian, S.A. Langer, A.J. Liu and S.R. Nagel, Phys. Rev. Lett. **89**, 095703 (2002).
6. A. Metha et G.C. Barker, Phys. Rev. Lett. **67**, 394 (1991).
7. G. D'Anna, A. Barrat and V. Lorento, F. Nori, Nature **424**, 909 (2003).
8. A. Coniglio and M. Nicodemi, Physica A **296**, 451 (2001).
9. E.R. Nowak, J.B. Knight, E. Ben-Naim, H.M. Jaeger and S.R. Nagel, Phys. Rev. E **57** 1971 (1998).
10. O. Pouliquen, M. Belzons and M. Nicolas, cond-mat/0305659.
11. A. Fierro, M. Nicodemi and A. Coniglio, Europhys. Lett. **59**, 642-647 (2002).
12. E.Clément, "Granular packing under vibrations ", in " *Physics of dry granular media*" p. 585, Ed. J.P. Hovi, H. Herrmann and S. Luding (Kluwer Acad. Publisher, 1998).
13. P. Evesque et J. Rajchenbach, Phys. Rev. Lett. **61**, 44 (1989).
14. C. Laroche, S. Douady and S. Fauve, J. Phys. (Paris) **50** 699 (1989).
15. E. Clément, J. Duran and J. Rajchenbach, Phys. Rev. Lett. **69**, 1189 (1992).
16. G. Ovarlez, C. Fond, E. Clément, Phys. Rev. E **67**, 60302 (2003).
17. J. Duran, T. Mazozi, E. Clément, J. Rajchenbach, Phys. Rev. E. **50**, 3092 (1994).
18. P. Philippe and D. Bideau, Phys. Rev. Lett. **91** 104302 (2003).
19. P. Philippe and D. Bideau, Europhys. Lett. **60**, 677 (2002).
20. J.B. Knight, C.G. Fandrich, C.N. Lau, H.M. Jaeger and S.R. Nagel, Phys. Rev. E **51**, 3957 (1995).
21. J.B. Knight, E.E. Ehrichs, V.Y. Kuperman, J.K. Flint, H.M. Jaeger and S. Nagel, Phys. Rev. E **54**, 5726 (1996).
22. D.A. Weitz and D.J. Pine. Diffusing-wave spectroscopy. In Wyn Brown, editor, *Dynamic light scattering: The method and some applications*, volume 49 of *Monographs on the physics and chemistry of material*, pages 652-720. Oxford University Press, Oxford, 1993.
23. A. Kabla and G. Debrégeas, cond-mat/0303560.
24. L. Cipelletti, H. Bissig, V. Trappe, P. Ballesta and S.J., Phys.:Condens Matter **15**, S257, (2003).
25. A. Metha and J.M. Luck, *Shaken not stirred: why gravel packs better than bricks*, present proceedings.
26. J.M. Luck and A. Metha, European Journal of Physics B, **35**, 399 (2003)
27. A. Crisanti and F. Ritort, cond-mat/0307554.

Unifying Concepts in Granular Media and Glasses
A. Coniglio, A. Fierro, H.J. Herrmann and M. Nicodemi (editors)

An oscillator in the granular matter

G. D'Anna[a] and P. Mayor[a]

[a]Groupe de systèmes complexes, Faculté des Sciences de Base, Ecole Polytechnique Fédérale de Lausanne, CH-1015 Lausanne, Switzerland

How close to a liquid is a "granular liquid"? Or a "granular glass"? To answer this question we have observed the motion of a torsion oscillator immersed in an externally shaken granular medium. We observe a Brownian-like behaviour in the granular liquid, with an approximate fluctuation-dissipation relation holding, and a Vogel-Fulcher-Tammann glass-like behaviour by decreasing the intensity of the external shaking, similar to the vitrification process of usual supercooled liquids.

1. INTRODUCTION

Under suitable external perturbation such as vibration or shear, a granular medium resembles an elastic solid, a gas, a liquid or even a glass[1]. How close is this resemblance? One way to provide an answer is to observe how a "tagged" grain behaves in, say, a granular liquid, and check if its motion resembles the Brownian motion which characterizes a tagged particle immersed in an usual liquid[2,3]. This Brownian motion results from the incessant thermal-induced tremor of the atomic or molecular-size particles composing the environment, and it has well known properties, captured for example in the Langevin model[4]. In a granular medium the thermal energy $k_B T$ at room temperature is too small to induce any macroscopic grain fluctuation, and a tagged grain would not move by itself. However, because of the very large number of degrees of freedom, one expects that an analogy to thermal-induced Brownian motion is possible when the system is externally driven and grain motion occurs by continuous injection of energy.

With this analogy in mind, we have developed a non-equilibrium version of a classical experiment, based on a torsion oscillator which is immersed into an externally vibro-fluidized granular system[5–7].

Should this oscillator have been immersed in a usual liquid at temperature T, the solution by Fourier transformation of the Langevin equation $I\ddot{\theta} + \alpha\dot{\theta} + G\theta = C_{ext}(t) + R(t)$ (where θ is the angular deflection, I is the moment of inertia, α the friction coefficient determining the viscous torque, G the fibre elastic torsion constant, C_{ext} the external torque, and $R(t)$ is the random torque, assumed to be a Gaussian white noise of noise strength q at correlation times larger than the microscopic relaxation time τ_x) would have given the noise power spectral density (PSD) $S(\omega) = 2q/I^2\left[\left(\omega_o^2 - \omega^2\right)^2 + \gamma^2\omega^2\right]$ where $\gamma = \alpha/I$ and the susceptibility $\chi(\omega) = \tilde{\theta}(\omega)/\tilde{C}_{ext}(\omega) = \chi'(\omega) - i\chi''(\omega)$, with $\chi(\omega) = 1/\left(\omega_o^2 - \omega^2\right)I + i\alpha\omega$. For a stationary process, the fluctuation-dissipation relation (FDR) would have linked the two parameters α and q, i.e., $S(\omega)\,\omega/\chi''(\omega) = 2q/\alpha = const.$

In thermodynamic equilibrium, the equipartition principle would have guaranteed that this constant be proportional to the temperature, in particular $2q/\alpha = 4k_BT$, which is the fluctuation-dissipation theorem (FDT). Notice that in the overdamped limit the Langevin equation reduces to the "minimal" form $\alpha\,\dot{\theta}= C_{ext}(t) + R\,(t)$, with a characteristic $1/f^2$ noise, i.e., $S\,(\omega) = 2q/\alpha^2\omega^2$. In this case the noise at a given frequency is proportional to the inverse microscopic relaxation time, i.e., $S(f) \propto 1/\tau_x$.

Should the oscillator have been immersed in an usual supercooled liquid, the FDT would have been violated, since the system, even if thermally driven, would not have been in thermodynamic equilibrium. However, a measure of the viscosity (or of the microscopic relaxation time, e.g., measuring the noise in the overdamped limit) then would have shown a slow-down of the dynamics, with a characteristic Vogel-Fulcher-Tammann (VFT) empirical law observed to link the noise and the temperature of the external bath.

By performing the non-equilibrium version of the experiment, that is by studying the susceptibility and the noise of the torsion oscillator immersed in a shaken granular medium, we have tested how far this Langevin approach can be used in the granular matter. Surprisingly, we find that the granular matter shares strong similarities with usual liquids at a given temperature, or even with supercooled liquids at low temperature. Here we summarize recent results[5–7].

2. EXPERIMENTAL

The granular material consists of glass beads of diameter d_g =1.1± 0.05 mm, and is contained in metallic cylinders of different sizes. We have obtained reproducible data using aged (i.e., exposed to atmosphere for a long time) glass beads. The probes are of different sorts: most of the data is obtained using a cone of apex angle 120°, or cylinders with different diameters. In general the probe is covered by a layer of glued beads. The design of the oscillator suspension system permits torsion motion, while other modes are strongly reduced. (Experimental details, such as the immersion depth or other geometrical parameters, can be found in[5–7,9]) We use two different oscillators, which have a natural frequencies f_n around 20 Hz, or around 12.5 Hz, depending on the suspension wires and the probe used. The first is used in the experiments in the liquid regime, and the second in the glass regime below.

The external perturbation is obtained using a vertical shaker, fed with signals of different type. We will thus generate different types of perturbation, such as: a high-frequency filtered white noise vibration, of central frequency f_c and band-with BW; a sinusoidal vibration of frequency f_s and amplitude a_s; a train of taps, separated by a time t_w, and where a single tap consists of one complete cycle of a f_s-sine wave. Notice that, except for the immersed probe, the oscillator is otherwise mechanically isolated from the container.

An accelerometer fixed on the container gives a measure of the acceleration spectrum, normalized to the acceleration of gravity, $A(\omega)$ (units g^2/Hz). We quantify the intensity of the external shaking by Γ, which is the square root of the band power of $A(\omega)$, in an extended band about 1 Hz to 10 kHz. In the case of a sinusoidal vibration of amplitude a_s and frequency $f_s = \omega_s/2\pi$, one recovers the usual definition of the vibration intensity in granular literature, i.e., $\Gamma = a_s\omega_s^2/g$. For sinusoidal vibration, $\Gamma = 1$ is the lift-off threshold, above which a single grain starts to fly.

3. GRANULAR LIQUID

Under relatively strong external perturbation we are able to establish a steady-state liquid regime, and measure the noise and the susceptibility, as shown in Fig. 1 for various vibration intensities Γ. In this experiment[7], the container is shaken with a high-frequency filtered white noise, of central frequency $f_c = 600$ Hz or 800 Hz and $BW = 600\ Hz$. The use of a filtered white noise is motivated by the wish to discard undesired effects such as pattern formation, rolls, and other sorts of instabilities in the granular motion[8]. The purpose is not to provide an ab initio random torque with white noise spectrum to the oscillator, since we observe the motion of the oscillator in a low-frequency range (10 Hz to 50 Hz) compared to the applied vibration high-frequency range.

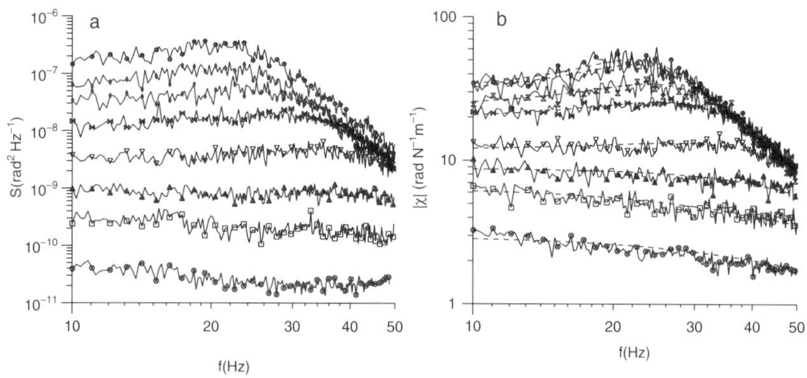

Figure 1. a) Noise PSD, $S(f)$, versus the frequency f, for different vibration intensities Γ; b) Modulus of the complex susceptibility $|\chi(f)|$ versus the frequency f, for different vibration intensities Γ. The intensity values are from top to bottom, 11.6, 10.0, 7.3, 5.4, 3.7, 2.2, 1.5 and 1.0.

The modulus of the complex susceptibility is fitted remarkably well by the susceptibility expression for the damped oscillator in usual thermal systems, and we thus can obtain the granular friction coefficient: we find systematically $\alpha \propto 1/\Gamma$. The FD-ratio $S(\omega)\omega/4\chi''(\omega)$ versus $f = \omega/2\pi$ at different Γ, is shown in Fig. 2. The FD-ratio is approximately independent of the frequency in the low-frequency range, providing evidence that the FD-ratio can be considered a constant in first approximation. The FD-ratio level can thus be used to define, by abuse of language, an *effective temperature*, T_{eff}. Plotting the averaged FD-ratio levels, i.e. $k_B T_{eff}$, versus Γ, we obtain $T_{eff} \propto \Gamma^2$. Notice that, since $\alpha \propto 1/\Gamma$ and $T_{eff} \propto \Gamma^2$, the granular friction coefficient α decreases by increasing the effective temperature, which is the behavior of liquids rather than gases.

This effective temperature is independent of the moment of inertia of the oscillator, or

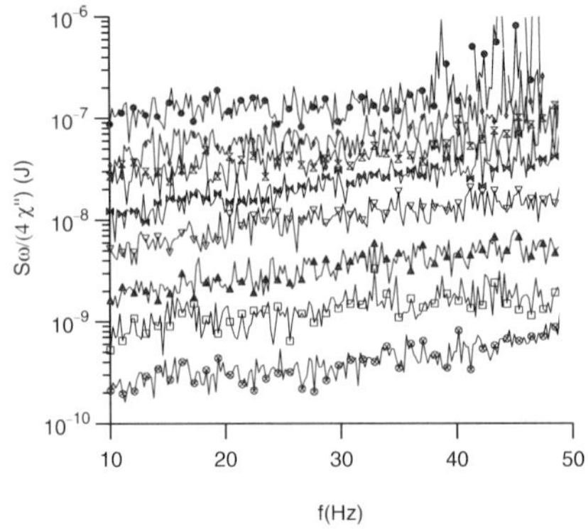

Figure 2. The FD-ratio versus Γ, extracted from the data of Fig. 2.

the diameter of the cylindrical probes. However, we observe that it depends on the angle of inclination of the surface exposed to the granular agitation, relative to the vertical axis: accordingly, we conclude that the effective temperature is an intrinsic parameter, but anisotropic, corresponding to horizontal and vertical impulse components. The effective temperature depends also on the immersion depth, and increases by increasing the immersion depth. A different behavior (still not understood) is seen very close to the free surface[7].

Although there is a small frequency dependence in the FD-ratio, the present experiment shows that one can satisfy the requirements for the Langevin approach to be applicable in a dissipative environment such as a vibrated granular medium.

4. GRANULAR GLASS

Weaker vibration corresponds to a lower effective temperature, and by analogy to a usual liquid one expects that by progressively reducing the shaken intensity the granular liquid should get "jammed" into a disordered configuration, following a process similar to the vitrification of supercooled liquids. In this "granular glass" (by abuse of language), the macroscopic dynamics is controlled by structural grain rearrangements. In particular, if τ_x represents the typical timescale between grain rearrangements, one expects a rapid increase of τ_x , i.e., a slow down of the dynamics.

The experiments[5,6] in the granular glass are more difficult than in the granular liquid, because of the overlapping of two effects: First, the probe becomes itself the origin of the external perturbation, since it moves up and down with respect to the container. Thus, sinusoidal vibrations are used, and the vibration intensity $\Gamma = a_s\omega_s^2/g$ must now be seen just a measure of the displacement a_s of the probe. Second, the inertia of the oscillator can fluidize the granular medium unless the dissipation is large enough. Thus, one needs to reach the overdamped limit, and we do this by increasing the immersion depth of the

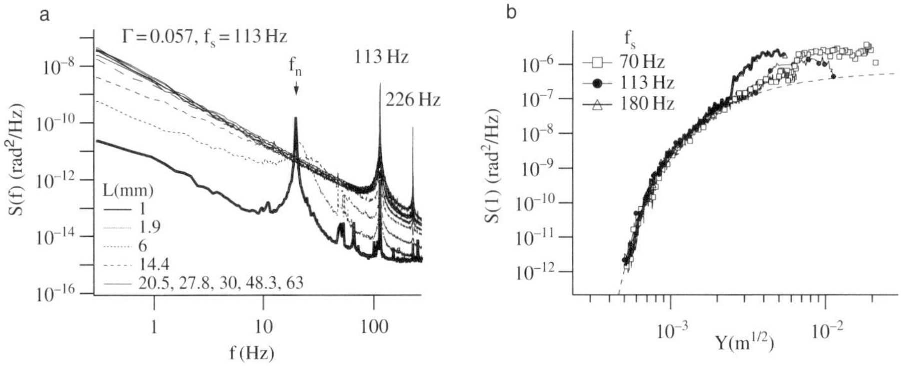

Figure 3. a) Noise PSD obtained by continuous vibration at 113 Hz, for different immersion depths L; b) The noise level S(1) versus the parameter $Y \equiv g^{1/2}\Gamma^{1/2}/2\pi f_s = \sqrt{a_s}$, for different f_s. The dashed line is a fit to a VFT -like form $S(1)| = C\exp[D(Y - Y_0)^p]$, with $p \approx -1$.

oscillator. Notice that in the granular glass we have not measured the susceptibility, but only the noise.

By increasing the immersion depth the noise becomes $1/f^2$, as shown in Fig. 3a, as expected in the overdamped limit. Accordingly, the oscillator becomes an elastically bounded random walker, which strictly "follows" the microscopic dynamics, and $S(f) \propto 1/\tau_x$. This provides an experimental way to measure τ_x. On the other hand, by changing the frequency of the external sinusoidal vibration, or using trains of taps, thus changing the up and down motion of the probe, we have found that the relevant control parameter is the square root of the displacement of the probe, $\sqrt{a_s}$. The noise at 1 Hz as a function of this parameter (for a large immersion depth, i.e., when the oscillator is overdamped) is shown in Fig. 3b, obtained using sinusoidal vibration of different frequencies. The data are fitted by a Vogel-Fulcher-Tammann (VFT) form, analogue to the one seen in the vitrification process of usual supercooled liquids.

5. DISCUSSION

There is a profound analogy between usual molecular matter subject to thermal motion, and granular matter subject to external vibration. Although in basic courses on thermal physics this analogy is vastly used in pictorial form, the present experiments shows that it is much deeper.

The oscillator, or the "fluctuation-dissipation thermometer", has shown that it make sense to introduce an effective temperature in strongly shaken granular matter. Apparently, the granular environment is mixing the relevant degrees of freedom fast enough with respect to viscous drag or diffusion, such as the use of the Langevin approach is justified, although the anisotropy and inhomogeneity of the medium must be taken into account. The observed behavior of the granular viscosity versus the effective temperature (at the shaken intensity used here) is similar to the one of usual liquids, hence the denomination granular liquid instead of granular gas.

As adequate vibration generates a granular liquid, by decreasing the vibration intensity (thus cooling down) a supercooled granular liquid (or simply a "granular glass") shows up[10]. In fact, the oscillator sees a noise with VFT like dependence between the noise level, or inverse microscopic relaxation time, and the intensity of the external shaking. A similar behavior is also seen in experiments aimed at studying the slow compaction dynamics of a granular medium submitted to separate vertical taps[11–13].

The grain level mechanics in the granular glass is very complicated, as deduced from the oscillator noise[14] or from direct observation[15], namely the distinction of a large number of localized grain fluctuations (inter-cage motion) and rare grain motion events that cause macroscopic rearrangement of the entire granular medium. The distinction between these "mesoscopic" grain fluctuations and the emergence of some rare, extreme events at the macroscopic level, might be a keen point for the understanding of slow dynamics in usual thermal systems, too. Finally, the measurement of the effective temperature in the granular glass using the oscillator, in particular studying the development or absence of development of a frequency (or time-) dependence of the FDR, can possibly bring together the recent theoretical developments in this field[16–18].

REFERENCES

1. H.M. Jaeger, S.R. Nagel and R.P. Behringer, Rev. Mod. Phys. 68 (1996) 1259.
2. A. Einstein, Ann. Physik 17 (1905) 549.
3. R. Brown, Phil.Mag. 4 (1828) 161.
4. R. Kubo, M. Toda and N. Hashitsume, Statistical Physics II (Springer, Berlin, 1991).
5. G. D'Anna and G. Gremaud, Nature 413 (2001) 407.
6. G. D'Anna and G. Gremaud, Phys. Rev. Lett. 87 (2001) 254.
7. G. D'Anna, P. Mayor, A. Barrat, V. Loreto and F. Nori, Nature 424 (2003) 909.
8. F. Melo, P.B. Umbanhowar and H.L. Swinney, Phys. Rev. Lett. 75 (1995) 3838.
9. G. D'Anna, Phys. Rev. E 62 (2000) 982.
10. A. Coniglio, A. de Candia, A. Fierro, M. Nicodemi and M. Tarzia, cond-mat/0309612.
11. J.B. Knight et al., Phys. Rev. E 51(1995) 3957.
12. C.S. O'Hern, L.E. Silbert, A.J. Liu and S.R. Nagel, Phys. Rev. E 68 (2003) 011306.
13. P. Philippe and D. Bideau, Phys. Rev. Lett. 91 (2003) 104302.
14. G. D'Anna et al., Europhys. Lett. 61(2003) 60.
15. O. Pouliquen, M. Belzons and M. Nicolas, Phys. Rev. Lett. 91 (2003) 014301.
16. A. Mehta and S.F. Edwards, Physica A 157 (1989) 1091.
17. L.F. Cugliandolo, J. Kurchan and L. Peliti, Phys. Rev. E 55 (1997) 3898.
18. V. Colizza, A. Barrat and V. Loreto, Phys. Rev. E 65 (2002) R050301.

Unifying Concepts in Granular Media and Glasses
A. Coniglio, A. Fierro, H.J. Herrmann and M. Nicodemi (editors)

Segregation phases in a vibrated binary granular layer

P.M. Reis[a*], T. Mullin[a] and G. Ehrhardt[b]

[a]Manchester Center for Nonlinear Dynamics, Department of Physics and Astronomy, University of Manchester, Oxford Road, Manchester M13 9PL, UK

[b]Theoretical Physics Group, Department of Physics and Astronomy, University of Manchester, Oxford Road, Manchester M13 9PL, UK

We present the results of an experimental study of patterned segregation in a horizontally shaken shallow layer of a binary mixture of dry particles. As the compacity, C, of the mixture was increased, the evolution of three distinct phases was observed. We classify them as binary gas, segregation liquid and segregation crystal phases using macroscopic and microscopic measures. The binary gas to segregation liquid transition is consistent with a continuous phase transition and includes the characteristic feature of critical slowing down. At high compacities we observed an intriguing slow oscillatory state.

1. Introduction

In segregation, excitation via flow or shaking can, counter-intuitively, cause an initially homogeneous mixture of grains to de-mix [1, 2]. Intriguingly, it does not always happen and the conditions for its occurrence are difficult to predict. Hence, a better understanding would have a major economic impact in the pharmaceutical, chemical processing and civil engineering industries [3]. However, despite decades of research, a predictive model of the process has yet to emerge. More recently, segregation has received considerable attention from the physics community as an example of a challenging far from equilibrium system with pattern formation [2]. Although many small scale laboratory studies involving vibration [4], avalanching in partially filled horizontal rotating drums [5] and stratification in vertically poured mixtures [6] have been investigated and a variety of mechanisms proposed [7], an understanding of the fundamental principles involved remains incomplete.

In our experiments [8, 1, 9, 10] we have focused on quasi-2-dimensional layers of binary mixtures of particles which are horizontally driven by the stick–slip frictional interaction with the surface of a horizontal tray. Vibration of an initially homogeneous mixed binary layer can give rise to robust segregation of the constituents, depending on the total filling fraction of the layer. We call this parameter the *compacity.* As the compacity was increased the evolution of distinct phases was observed. We identified them as a *binary gas*, a *segregation liquid* and a *segregation crystal* using both microscopic and macroscopic measures to identify their properties. This horizontal and quasi-2-dimensional set up

*email address: pedro@reynolds.ph.man.ac.uk

gives the practical advantage that any collective behavior is readily visualized, vertical compaction effects are effectively eliminated and the material is in contact with the forcing throughout the drive cycle.

For this class of quasi-2D binary granular systems a qualitative segregation mechanism has been suggested [10, 11, 12, 13] using the idea of *excluded volume depletion* as in colloidal systems and binary alloys [15]. This is in the spirit of the physics of complex systems where attempts have been made to unify descriptions of granular materials, colloids, gels and foams [17, 18]. We discuss the relevance of these ideas to our observations in some concluding remarks.

2. The experiment

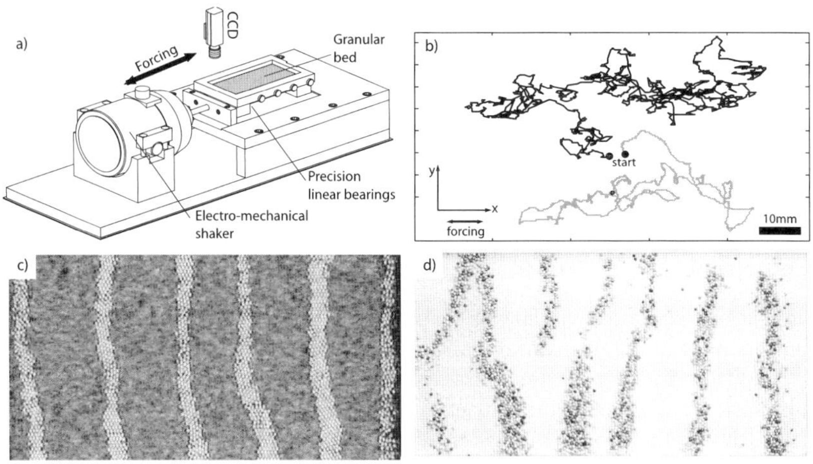

Figure 1. (a) Schematic diagram of the experimental apparatus. (b) Trajectories of two individual particles over $1min$: black line corresponds to poppy seed and grey line to phosphor-bronze sphere. (c) Typical segregation pattern for mixture A: poppy seeds (grey regions) and phosphor-bronze spheres (white regions). (d) Typical segregation pattern for mixture B: sugar particles (shaded grey regions) and polystyrene spheres (white regions). Both patterns self-organised $15min$ after starting from a homogeneous mixture.

A schematic diagram of the top view of the apparatus is presented in Fig. 1(a). It consisted of an accurately levelled and horizontal smooth rectangular tray, of dimensions $(x, y) = 180 \times 90mm$ with a flatness of less than $\pm 5\mu m$, on which particles were vibrated longitudinally. The tray was made from aluminum tool plate for increased rigidity and was mounted on a horizontal platform which was connected to an electro-mechanical shaker. Its motion was a close approximation to one-dimensional as it was constrained by four lateral high precision linear bush bearings. The dynamic displacement and acceleration of

Material	Avg. Diameter (mm)	Density (gcm^{-3})	Shape
poppy seed	1.07 (±17%)	0.2	flat 'kidney' shaped
phosphor-bronze	1.50	8.8	precision spherical
polystyrene	0.5	1.1	precision spherical
sugar	1.71	1.6	roughly spherical

Table 1
Physical properties of the particles used.

the shaking bed was monitored by a Linear Displacement Variable Transformer (LVDT) and a piezoelectric accelerometer. The tray's acceleration was checked to be sinusoidal, to 0.1% in power, using the power spectrum of the accelerometer trace.

We have investigated segregation using a variety of mixtures but here focus on results for two specific types: *mixture A* (polystyrene spheres + sugar particles) and *mixture B* (phosphor-bronze spheres + poppy seeds). In particular, we have used mixture B for the detailed quantitative results. The details for the particles' material properties is presented in Table 1.

At this point we define the *layer compacity* to be the total filling fraction of the system,

$$C = \frac{N_1 A_1 + N_2 A_2}{xy}, \tag{1}$$

where N_1 and N_2 are the numbers of species a and b in the layer, A_1 and A_2 are the two dimensional projected areas of the respective individual particles ($A_1 = (0.87 \pm 0.15)mm^2$ for poppy seeds and $A_2 = (1.77 \pm 0.06)mm^2$ for phosphor-bronze spheres) and x and y are the longitudinal and transverse dimensions of the rectangular tray. We have chosen to fix the forcing at a single frequency $f = 12Hz$ and amplitude $A = \pm 1.74mm$ and explore the effect of the layer compacity.

Throughout this study, experiments were performed in a quasi-monolayer regime so that the smaller particles were rarely on top of the larger ones. This was particularly the case for the mixtures of poppy seeds and phosphor bronze spheres since the size ratio for these two types of particle is $\sim 2:3$. However, some overlapping of particles did occur so that compacities up to $C \sim 1.1$ were achieved in the high packing limit.

3. Single particle motion

As discussed above, the forcing was strictly sinusoidal but both types of particles were driven by a stick-and-slip interaction with the oscillatory surface of the container. These interactions induced stochastic motion and trajectories of single particles with the characteristics of a 2D random walk were obtained. Typical one minute long trajectories for a poppy seed and a phosphor-bronze sphere are shown in Fig. 1(b). The statistics of the motion were anisotropic with a preference for motion in the direction of the forcing. Note that the two types of particles responded to the driving differently, due to their mass, shape and surface properties. When many particles were present, collective motion dominated with diffusion biased in the direction of the forcing.

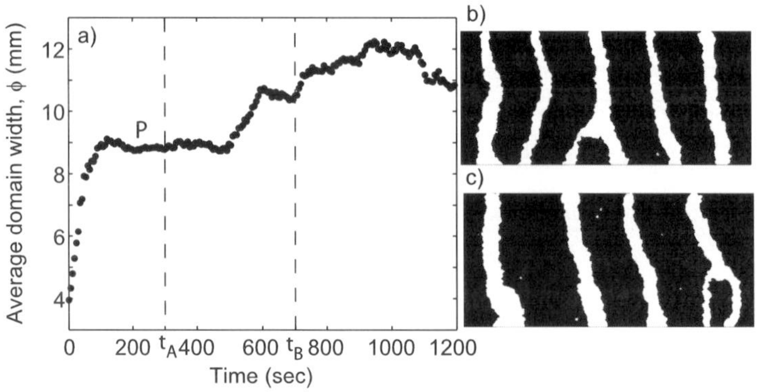

Figure 2. (a) Time series of the average domain width, ϕ, over 20min, for $C = 1.057$. (b)&(c) Binary frames of the segregation patterns (regions of poppy seeds in black and regions of phosphor-bronze spheres in white) at (b) $t_A = 300sec$ and (c) $t_B = 702sec$.

4. Granular segregation

In Fig. 1(c) and (d) we present two examples of segregation patterns for the two mixtures considered. The snapshots correspond to the segregated states which formed after the mixture was vibrated for a period of 15 minutes. The experimental runs were systematically started from a homogeneous mixed layer using the following method. Firstly, N_2 small particles were vibrated at large amplitudes, $A \sim \pm 5mm$, creating an homogeneous and isotropic sub-monolayer. The large particles were then suspended above the layer, on a horizontal perforated plate with (57×28) $2mm$ diameter holes arranged in a triangular lattice. A shutter was then opened and the 1596 particles fell onto the layer of small particles, creating a near homogeneous binary mixture. Domains such as those shown in Fig. 1(c) and (d) then formed. The differences in the physical properties of the two mixtures illustrates that, in our system, patterned segregation is a robust phenomena.

In order to quantify the dynamics of growth of the segregated domains we focus on mixture B and establish two macroscopic measures: the average longitudinal width and the average number of domains of phosphor-bronze spheres. At this stage, the images of the whole layer $(300 \times 600$ pixels) were processed such that regions of poppy seeds were given a value of 0 (black) and regions of spheres a value of 1 (white). For each acquired frame this was calculated by scanning through each of the 300 horizontal lines and the length, L_i, of each of the *white steps* was measured. We therefore define, $\phi = \langle L \rangle$, to be the average value of the domain width and the average number of domains, η, to be the number of such steps, per pixel line.

A typical time evolution of ϕ for $C = 1.057$, over a period of 20min, is presented in Fig. 2(a). It is characterised by a fast initial segregation growth, over timescales of the order of $\sim 1min$. During this period the mechanism of the segregation process is as follows. Initially, single large particles diffuse in a sea of the smaller ones. When two large particles

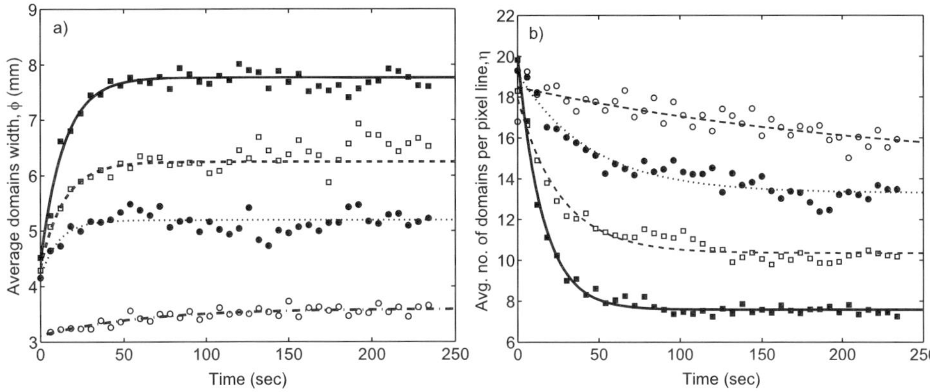

Figure 3. (a) Time-series of the average domain width, ϕ. (b) Time-series for number of domains, η. (□) $C = 0.873$, (■) $C = 0.729$, (•) $C = 0.708$, (○) $C = 0.602$. Lines are fits to Eqn. (2). All runs were started from a homogeneously mixed layer.

come close together, the smaller particles cannot fit between them, and hence the pair is subjected to an asymmetric pressure that will keep it together. Subsequently, pairs may encounter others so that progressively larger clusters form. The unidirectionality of the driving induces an asymmetry in the segregated domains such that elongated domains of the larger particles (phosphor-bronze spheres or sugar particles) develop. The segregation domains were elongated in a direction perpendicular to the drive. The growth of the domains then exhibits an intermediate saturation level as can be seen in the plateau labelled P in Fig. 2(a).

Further coarsening of the domains can take place, albeit at longer timescales. Two examples of the patterns at later times are presented in Fig. 2(b) and (c) at $t_A = 300s$ and $t_B = 702s$, respectively. During this period coarsening has occurred via the break-up of one of the stripes and consequent merging with its neighbours. In contrast to the fast initial segregation, this slow coarsening may be associated with large scale collective motion in the layer of poppy seeds.

Since there is a distinct separation of the timescales we will now focus on the *fast* initial segregation growth regime. In Fig. 3(a) and (b), we present typical time-series for ϕ and η, respectively, for four different values of C, over the first $4min$ of vibration. The superimposed lines are fits of,

$$\phi(C,t) = \Sigma^{\phi} - b^{\phi}.\exp\left(-\frac{t}{t_s^{\phi}}\right) \quad \text{and} \quad \eta(C,t) = \Sigma^{\eta} + b^{\eta}.\exp\left(-\frac{t}{t_s^{\eta}}\right), \qquad (2)$$

to the experimental time series of both ϕ and η, where Σ^{ϕ} and Σ^{η} are the respective saturation levels and t^{ϕ} and t_s^{η}, the *segregation times*, are timescales associated with the saturation of the domain growth. We denote Σ^{ϕ} by the *segregation level.* A result which may be inferred from both sets of time sequences in Fig. 3 is that, when the domains

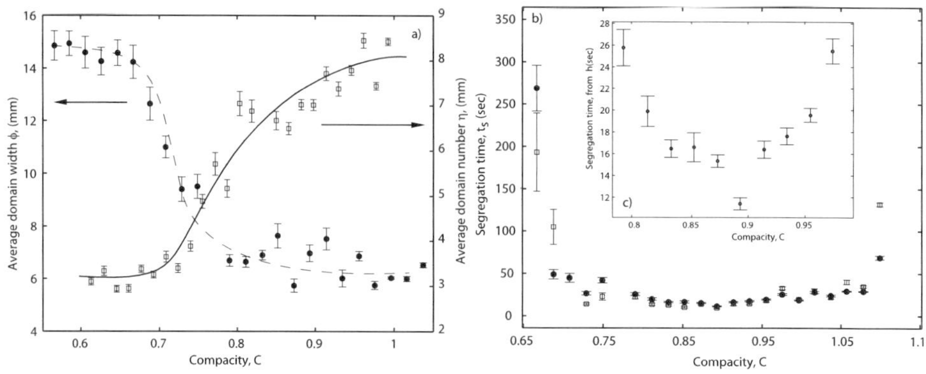

Figure 4. Segregation phase transition. (a) Compacity dependence of saturation levels from average domain width (•) and average number of domains (□). (b) Segregation time, t_s. from average domain width (□) and average number of domains (•). (c) Inset: Segregation time from average number of domains zoomed around $C = 0.893$. Lines are guides to the eye.

form, there is a progressive coalescence of clusters with an increase of the average width and corresponding decrease in the number of domains in the system.

5. Segregation Phase transition

We next investigated the dependence of the saturation levels, Σ, and segregation times, t_s, on the compacity of the layer for both ϕ and η. This was done by incrementally increasing the number of poppy seeds, N_1, in the layer, in measured steps, while the number of phosphor-bronze spheres, $N_2 = 1596$, held constant, i.e. $C = C(N_1)$. We chose to perform the experiments by changing the numbers of poppy seeds. This can be regarded analogous to the experimental protocol in colloid-polymer mixtures of increasing the concentration of the polymer in solution (see Section 8 and Ref. [21]). The qualitative behaviour presented here was found to be robust over a range of N_2.

The C-dependence of the segregation level is presented in Fig. 4(a). For low compacities, up to a value, $C_c \sim 0.647$, the segregation level remains low and approximately constant around $(3.25 \pm 0.24)mm$. This corresponds to approximately two sphere diameters and we do not classify these as domains; the system is deemed to be in a mixed or unsegregated state. This is essentially a collisional regime since there is enough free volume in the system so that particles collide and randomly diffuse, without aggregation. We denote this regime by a *binary gas*.

As C is increased, past C_c, clusters of the larger particles form mobile segregation domains, of increasing ϕ, emerge in a nonlinear way. Above C_c, the clusters of larger particles were mobile and their movement was reminiscent of oil drops on water as they flowed, merged and split and the motion of the particles within the clusters was highly agitated. As C was incrementally increased there was a gradual decrease of the overall motion of segregation domains so that they become less mobile and increasingly more stable. Also, for larger values of C, there was a significant decrease of the agitation of the spheres within the domains.

The average number of domains, shown in Fig. 4 remains approximately constant up to the same value C_c, after which it rapidly decreases as C is increased. This is expected for increasingly wider domains since the number of phosphor-bronze spheres is constant. This scenario is consistent with the existence of a continuous phase transition for granular segregation of the binary mixture, as the layer compacity is incrementally increased, with a critical point for segregation, $C_c \simeq 0.647$.

In Fig. 4(b) we present the segregation times obtained from the fits of Eq. (2) to $\phi(t)$ and $\eta(t)$ as a function of C. Note that we only define t_s for $C > C_c$, since this is the parameter range over which segregation occurs. As expected for a continuous phase transition, t_s diverges, as C is decreased from above, near C_c, a phenomena usually referred to as *critical slowing down*[14]. This adds further evidence for the existence of a phase transition.

As the compacity is increased, past C_c, this slowing down of the segregation persists up to $C = 0.893$. At this point there is a minimum in the segregation time, as shown in Fig. 3(c), with $t_s = 11.41s$. For greater values of C there is a qualitative change in the segregation behaviour and the layer takes increasingly longer to segregate. At this large values of C, stable crystalline stripes form which are separated by dense regions of poppy seeds.

The above scenario suggests that the segregated phase, for $C > C_c$, can be sub-divided into two further regimes: one in which the speed of segregation increases with C ($0.647 < C < 0.893$) and another in which the opposite occurs ($C > 0.893$). We denote the first by *segregation liquid* and the second by *segregation crystal.*

6. Binary gas, segregation liquid and segregation crystal phases

To further justify this idea of three segregation phases we turn to the structural information given by the radial pair distribution function for the large spheres, $g(r)$,

$$g(r) = A(r)\langle\sum_i\sum_{j\neq i}\delta(r - r_{ij})\rangle, \tag{3}$$

where r_{ij} is the separation of the i and jth particles and the angled brackets denote a time average. $A(r)$ is a normalization constant such that $g(r) = 1$ for a uniform distribution of particles. The data for the calculation of $g(r)$ was obtained 3min after the driving was switched on. Time averaging was then performed over 1min, for 1500 temporal realisations of the layer's configuration.

In Fig. 5(a) we show three curves of $g(r)$ where each represents the segregation phases discussed above. Typical configurations of the spheres, for each of the phases, are presented in the zoomed photographs in Fig. 5(b),(c) and (d). Note that the photographs have been image processed such that only the phosphor-bronze spheres are visible. For the binary gas ($C < 0.647$), $g(r)$ has a peak at $r/d = 1$ which quickly decays at large distances, as expected for a disordered gas. Increasing C results in a monotonic increase of the height of the first peak, $g(d)$. At intermediate compacities ($0.647 < C < 0.893$), *liquid-like* behavior is observed, with $g(r)$ peaked at 1, 2 and 3 particle diameters. The positions

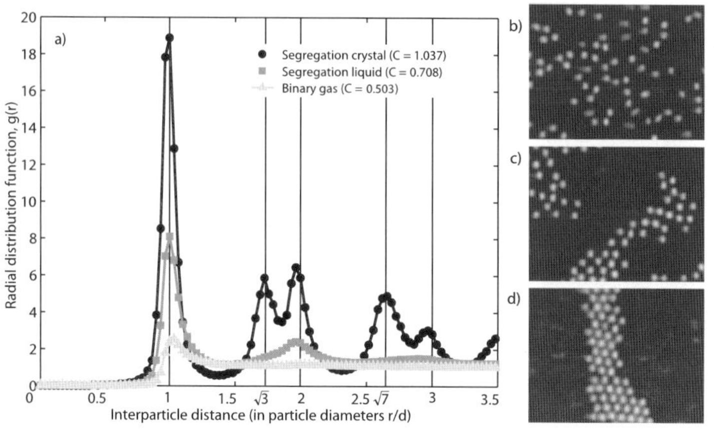

Figure 5. (a) Radial distribution functions, $g(r)$, for the phosphor-bronze spheres: (▲) binary gas (C=0.503), (■) segregation liquid (C=0.708) and (•) segregation crystal (C=1.037). The vertical dashed lines correspond to the expected location of the peaks for an hexagonal lattice in 2D. (b) Zoomed binary gas. (c) Zoomed segregation liquid. (d) Zoomed segregation crystal. The photographs have been image processed such that only the phosphor-bronze spheres are visible.

of neighbouring spheres are correlated and the maxima may be associated with shells of neighbours. The oscillations are rapidly damped, showing the decay of short-range order. This behaviour is commonly seen in hard sphere liquids and was first observed experimentally by Bernal [16]. In the segregation crystal phase ($C > 0.893$) two further peaks emerge near $r/d = \sqrt{3}$ and $r/d = \sqrt{7}$, characteristic of a hexagonally packed crystal, in two dimensions. At these high compacities, individual spheres in the segregation domains rattle inside the cage formed by its neighbours.

7. Oscillatory states

At the highest values of the compacity where robust crystalline stripes form, we have observed intriguing well defined oscillations of the segregation patterns. The phosphor-bronze stripes bend, periodically, backwards and forwards in a reproducible way. For each of the regions of poppy seeds in between two phosphor-bronze stripes, there is a complementary collective motion in the form of a single large vortex. The direction of rotation of the vortex is reversed every half-cycle of the oscillation. In Fig. 6(a)-(c) we show three photographs of the granular layer which were captured 40 min after having started from a homogenous mixture and correspond to the beginning of the cycle, half-cycle and full cycle points of one oscillation, respectively.

We illustrate the periodic nature of the oscillations, in Fig. 6(b) with a space time diagram of the process. This was constructed by sampling along a single line in the x-dimension, and progressively stacking the lines over a period of $1.5hrs$. The sampling

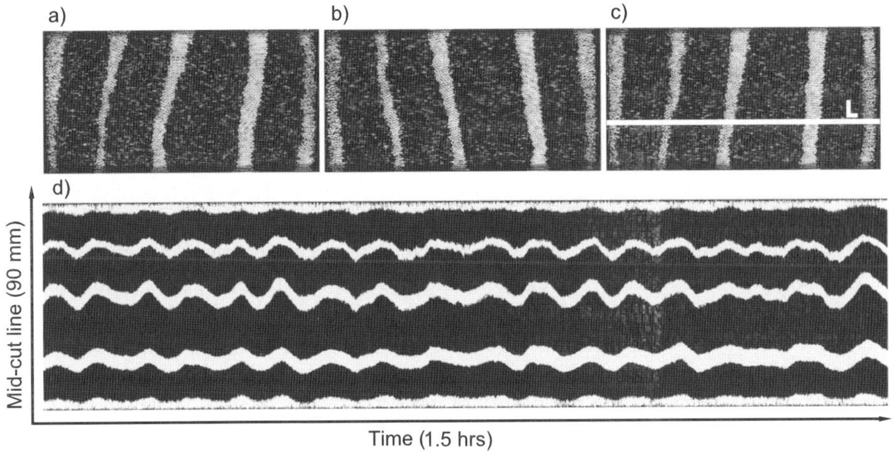

Figure 6. Photographs of the granular layer showing the oscillatory bending of the phosphor-bronze stripes (regions in white), at different times: (a) beginning of cycle, (b) half-cycle, (c) end of cycle. (d) Space time diagram constructed from stacks of the cut line, L in (c), over $1.5hrs$.

line was positioned at 1/3 of the y-length of the tray as indicated by the solid white line superposed in Fig. 6(c). A well defined cycle is clearly observed and it has a period of $T = 286s$. This suggests the existence of a strong coupling between the oscillation of the stripes through bending and the vortex inversion in the poppy seed's layer. It is interesting to note that the timescale associated with these oscillations is $\sim$ 3500 times slower than the driving (at $12Hz$) excluding the possibility of a simple resonance in the driving/granular-layer system. Hence, the mechanism which gives rise to this oscillatory phenomena remains, as yet, unexplained.

8. Discussion and conclusion

It is interesting to note that there are some analogies between segregation in our binary granular system and aggregation in binary colloids [19] and emulsions [20]. In particular, when the size ratio between the large colloidal particles and the radius of gyration of the polymer is larger than $\sim$ 0.3, colloid-polymer mixtures display colloidal gas, liquid and crystal phases of the large spheres [21, 22]. Aggregation in such systems is usually explained using an excluded volume depletion argument introduced by Asakura and Oosawa [23] in the context of binary hard spheres. This entropic argument is equivalent to the mechanistic view that if two large particles are close enough so that no other particle (or polymer) may fit between them, they will be subjected to an asymmetric osmotic pressure that leads to an effective attractive inter-particle force. In colloid-polymer mixtures the strength of the interaction can be tuned by changing the concentration of the polymer in solution, therefore inducing gas-to-liquid and liquid-to-crystal transitions. These resemble the granular phases we have observed by incremental increase of the compacity. However, an important difference is that our granular system is both driven and dissipative and therefore far from equilibrium. In addition, Brownian motion is an intrinsic part of the equilibrium dynamics of colloidal systems. Hence a direct connection between the two systems remains an open question.

In conclusion, our results provide quantitative evidence for the existence of three phases - binary gas, segregation liquid and segregation crystal - in granular segregation of horizontally excited binary mixtures. At high compacities, intriguing regular oscillatory states were observed. We show that segregation undergoes a continuous phase transition that occurs for compacities above a critical value and exhibits critical slowing down. We have presented both macroscopic and microscopic measures which are self-consistent. Analogies of this phase behaviour may be drawn with other depletion driven self-assembling binary systems. This raises the possibility of importing ideas from binary colloids in equilibrium to formulate new models for granular segregation.

The authors would like to thank A. Stephenson for valuable discussions. P.M.R. was supported by the Portuguese Foundation of Science and Technology. G.E. was funded by a studentship from the EPSRC. The research of T.M. is supported by an EPSRC Senior Fellowship. This work was completed while visiting the Isaac Newton Institute, Cambridge.

REFERENCES

1. T. Mullin. *Science*, 295:1851, 2002.
2. T. Shinbrot and F. Muzzio. *Physics Today*, pages 25–30, March 2000.
3. J. C. Williams. *Powder Technology*, 15:245–251, 1976.
4. N. Burtally, P. King, and M. Swift. *Science*, 295:1877–1879, 2002.
5. J. M. N. T. Gray. *J. Fluid. Mech.*, 441:1–29, 2001.
6. H. Makse, P. Cizeau, and H. Stanley. *Phys. Rev. Lett.*, 78:3298–3301, 1997.
7. J. M. Ottino and D. V. Khakhar. *Annu. Rev. Fluid Mech.*, 32:55–91, 2000.
8. T. Mullin. *Phys. Rev. Lett.*, 84(20):4741–4744, 2000.
9. P. M. Reis and T. Mullin. *Phys. Rev. Lett*, 89:244301, 2002.
10. P. M. Reis, G. Ehrarhdt, A. Stephenson and T. Mullin. submitted to *Europhys. Lett.*, August 2003.
11. S. Aumaitre, T. Schnautzand C. A. Kruelle, and I. Rehberg. *Phys. Rev. Lett.*, 90:114302, 2003.
12. T. Shinbrot and F. Muzzio. *Nature*, 410:251–258, March 2001.
13. J. Duran and R. Jullien. *Phys. Rev. Lett*, 80:3547–3550, 1999.
14. P. M. Chaikin. *Principles of condensed matter physics*, (CUP), 1995.
15. K. M. Hill and J. M. Kakalios. *Phys. Rev. E*, 49:R3610, 1994.
16. B.J. Bernal, *Proc. Roy. Soc*, A280:299, 1964.
17. A. J. Liu and S. R. Nagel. *Nature*, 396:21–22, 1998.
18. V. Trappe, V. Prasad, L. Cipelletti, P. N. Segre, and D. A. Weitz. *Nature*, 411:772–775, 2001.
19. A. D. Dinsmore, A. G. Yodh, and D. J. Pine. *Phys. Rev. E*, 52:4045–4057, 1995.
20. J. Bibette, D. Roux, and F. Nallet. *Phys. Rev. Lett.*, 65:2470–2473, 1990.
21. W. C. K. Poon. *J. Phys.: Condens. Matter*, 14:R859–R880, 2002.
22. A. Moussaïd, W.C.K. Poon, P.N. Pusey and M.F. Soliva, *Phil. Trans. R. Soc. Lond. A*, 359:897–907, 2001.
23. S Asakura and F Oosawa. *J. Polym. Sci*, 33:183–192, 1958.

Unifying Concepts in Granular Media and Glasses
A. Coniglio, A. Fierro, H.J. Herrmann and M. Nicodemi (editors)

Shaken, not stirred: why gravel packs better than bricks

Anita Mehta[a*] and J.M. Luck[b†]

[a]S.N. Bose National Centre for Basic Sciences, Block JD, Sector 3, Salt Lake, Calcutta 700098, India

[b]Service de Physique Théorique, URA 2306 of CNRS, CEA Saclay, 91191 Gif-sur-Yvette cedex, France

We explore the effect of shape – jagged vs. regular – in the jamming limit of very gently shaken packings. Our measure of shape ε is the void space occupied by a disordered grain; we show that depending on its number-theoretic nature, two generic behaviours are obtained. Thus, regularly shaped grains (rational ε) have ground states of perfect packing, which are irretrievably lost under zero-temperature shaking; the reverse is the case for jagged grains (irrational ε), where the ground state is only optimally packed, but entirely retrievable. At low temperatures, we find intermittency at the surface, which has recently been seen experimentally.

1. Introduction

Why does shaken gravel pack better than shaken bricks do? We explore this question via a one-dimensional model [1,2]; despite its simplicity, it exhibits frustration and slow dynamics, features [3] which link the fields of granular compaction [4,5] and glasses [6]. The central issue that we probe is the effect of granular shape; our main finding is that irregular/jagged and regular/smooth grain shapes have rather different consequences for compaction in the jamming limit.

That glasses or granular media do not have crystalline ground states is well known, as is the fact that their attempts to reach their ground states are 'jammed', i.e., hindered by long-range interactions. Our model is based on the following picture of jamming. In the absence of holes (grain-sized voids), the only way that granular media can compact is by grain reorientation. A disordered grain, in our model, 'wastes space'; that is, it occupies a net volume equal to its size, plus that of a *partial void* [1,2,7].[1] A reorientation of this to an ordered ('space-saving') state 'frees up' the partial void, for use by other grains to reorient themselves. This cascade-like picture of compaction has been seen in the comparisons with experiment [8] of random graph models [9] of granular compaction. Also, as there [9], the response of grains to external dynamics is the *local* minimisation

*anita@bose.res.in
†luck@spht.saclay.cea.fr
[1]Specifically, in our model, each ordered grain occupies one unit of space, and each disordered grain occupies $1+\varepsilon$ units of space, where ε is a measure of the trapped void space for a given granular shape.

of void space; the ground states so obtained [1,2] resemble much more the random close-packed state found in granular systems [10] than the rather unrealistic crystalline ground state obtained in earlier work [7].

2. The model: definition and ground states

Grains are indexed by their depth n, measured from the surface of the column. Each grain can be in one of two orientational states – ordered (+) or disordered (−) – the 'spin' variables $\{\sigma_n = \pm 1\}$ thus uniquely defining a configuration. We posit an ordering field h_n which constrains the temporal evolution of spin σ_n, such that the excess void space is minimised – a constraint which is reasonable in the jamming limit.

The stochastic dynamics in the presence of a vibration intensity Γ is defined by the transition probabilities:

$$w(\sigma_n = \pm \to \sigma_n = \mp) = \exp(-n/\xi_{\rm dyn} \mp h_n/\Gamma). \tag{1}$$

The dynamical length (or boundary layer [1,2]) $\xi_{\rm dyn}$ is a measure of the extent to which free surface effects percolate into the bulk; well beyond this, the dynamics is *slow*, while within it, the free surface still has an effect on the dynamics which are relatively fast. The local ordering field h_n reads

$$h_n = \varepsilon\, m_n^- - m_n^+, \tag{2}$$

where m_n^+ and m_n^- are respectively the numbers of + and − grains above grain n. Equation (2) shows that a transition from an ordered to a disordered state for grain n is *hindered* by the number of voids that are already above it: in fact h_n is a measure of the *excess void space* [11] in the system.

In the $\Gamma \to 0$ limit of *zero-temperature dynamics* [9], the probabilistic rules (1) become deterministic:

$$\sigma_n = \text{sign } h_n, \tag{3}$$

provided $h_n \neq 0$ (see below). Ground states are the static configurations obeying (3) everywhere. A rich ground-state structure is achieved for $\varepsilon > 0$, because of *frustration* [3], whose nature depends on whether ε is rational or irrational. We mention for completeness that the case $\varepsilon < 0$ is a generalisation of earlier work [7], with a complete absence of frustration and a single ground state of ordered grains.

The *rotation number* $\Omega = \varepsilon/(\varepsilon + 1)$ fixes the proportions of ordered and disordered grains in the ground states: $f_+ = \Omega$, $f_- = 1 - \Omega$. For irrational ε, (2) implies that all the local fields h_n are non-zero. Qualitatively, irrational values of ε denote shape irregularity; the above then implies that for such jagged grains, the excess void space is never zero, or the packing is never perfect, even in the ground state. It turns out [1,2] that the ground state is in fact *quasiperiodic*; the local fields h_n lie in a bounded interval $-1 \leq h_n \leq \varepsilon$.

For rational $\varepsilon = p/q$, with p and q mutual primes, $\Omega = p/(p+q)$, and some of the h_n can vanish. This means that grain n has a perfectly packed column above it, so that it is free to choose its orientation. It turns out [1,2] that orientational indeterminacy occurs at points of perfect packing such that n is a multiple of the *period* $p+q$.[2] Each ground state is thus a random sequence of two patterns of length $p+q$, each containing p ordered and q disordered grains. The model therefore has a *zero-temperature configurational entropy* or *ground-state entropy* $\Sigma = \ln 2/(p+q)$ per grain. Qualitatively, rational values of ε imply a regularity or 'smoothness' of grain shape; one could imagine that regular grains would align themselves to fit exactly into available voids where possible, in a ground state configuration. This in fact happens, leading to states of perfect packing at various points of the column and the observed degeneracy of ground states.

3. Zero-temperature dynamics: (ir)retrievability of ground states

Zero-temperature dynamics is a theoretical construct; one assumes that, starting with a random array of objects, the limit of zero shaking intensity will cause their ground state of packing to be achieved. This is neither obvious, nor, as we will show, correct in general.

Our analysis shows [1,2] that, under zero-temperature dynamics, jagged grains are able to retrieve their unique ground state, whereas for regular grains, the true ground states are impossible to retrieve. In the latter case, one finds instead a steady state with non-trivial *density fluctuations* above the ground states, which recall the observed density fluctuations above the random close-packed state [12,13] in real granular materials. This is intuitively understandable; jagged grains in the limit of extreme compaction can only 'click' into place in a unique way, while the huge degeneracy of perfectly packed ground states for regular grains makes any particular one impossible to retrieve, using a random shaking dynamics.

We recall the rule for zero-temperature dynamics:

$$\sigma_n \to \operatorname{sign} h_n. \tag{4}$$

For irrational ε with an initially disordered state, this results in the recovery and ballistic propagation of the quasiperiodic ground state, starting from the free surface to a depth $L(t) \approx V(\varepsilon)\,t$. The velocity $V(\varepsilon) = V(1/\varepsilon)$ varies smoothly with ε, and diverges as $V(\varepsilon) \sim \varepsilon$ for $\varepsilon \gg 1$ [1,2]. The rest of the system remains in its disordered initial state. When $L(t)$ becomes comparable with ξ_{dyn}, the effects of the free surface begin to be damped. In particular for $t \gg \xi_{\text{dyn}}/V(\varepsilon)$ we recover the logarithmic coarsening law $L(t) \approx \xi_{\text{dyn}} \ln t$, observed in related work [7,9] to model the slow dynamical relaxation of vibrated sand [12].

For rational ε, things are more complex; the local field h_n in (4) may vanish. We choose to update such orientations according to $\sigma_n \to \pm 1$ with probability $1/2$, leading to a dynamics which is stochastic even at *zero* temperature. Here, even the behaviour well within the boundary layer ξ_{dyn} contains many intriguing features, while the dynamics for $n \gg \xi_{\text{dyn}}$ is again logarithmically slow [1,2]. Focusing on the limit $\xi_{\text{dyn}} = \infty$, our main result is that *zero-temperature dynamics does not drive the system to any of its degenerate*

[2]For $\varepsilon = 1/2$, for example, one can visualise that each disordered grain 'carries' a void half its size, so that units of perfect packing must be permutations of the triad $+--$, where the two 'half' voids from each of the $-$ grains are filled by the $+$ grain. The dynamics, which is *stepwise compacting*, selects only two of these patterns, $+--$ and $-+-$. Evidently this is a one-dimensional interpretation of packing, so that the serial existence of two half voids and a grain should be interpreted as the insertion of a grain into a full void in higher dimensions.

ground states. The system instead shows a fast relaxation to a non-trivial steady state, independent of initial conditions. In this steady state, the local fields h_n have unbounded fluctuations as a function of depth [1,2]; since h_n is a measure of excess void space, this in turn implies the existence of *density fluctuations* in our model of shaken sand.

Fig. 1 shows the variation of these density fluctuations as a function of depth n:

$$W_n^2 = \langle h_n^2 \rangle \approx A\, n^{2/3}, \quad A \approx 0.83. \tag{5}$$

The fluctuations are approximately Gaussian, with a definite excess at *small* values: $|h_n| \sim 1 \ll W_n$. Interestingly, this 'nearly but not quite' Gaussian behaviour of density fluctuations in shaken granular media has been seen experimentally [8]; the not-quite-Gaussianness was in the experiment interpreted via correlations, which as we will show below, are also present in our model.

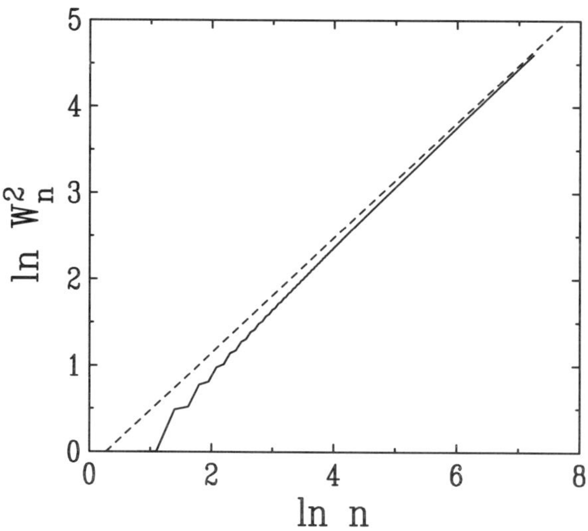

Figure 1. Log-log plot of $W_n^2 = \langle h_n^2 \rangle$ against depth n, for zero-temperature dynamics with $\varepsilon = 1$. Full line: numerical data. Dashed line: fit to asymptotic behaviour leading to (5) (after [1,2]).

To reiterate, (5) implies that the known ground states of a system of regularly shaped grains will never be retrieved even in the limit of zero-temperature dynamics. The steady state will, instead, be one of *density fluctuations* above the ground state. The present model, to our knowledge, thus contains the first derivation[3] of a possible source of density

[3] A simple scaling argument explains the observed roughening exponent 2/3. Let h_n be the position of a random walker at 'time' n. The noise in this fictitious random walk originates in the sites $m < n$ where the local field h_m vanishes. It is therefore proportional to $\sum_{m=1}^{n-1} \mathrm{Prob}\{h_m = 0\}$, hence the consistency condition $W_n^2 \sim \sum_{m=1}^{n-1} 1/W_m$, yielding the power law (5).

fluctuations in granular media [12,13], which, here, arise quite naturally from the effects of shape. This prediction of complex experimental observations [8,12] is all the more startling given its origin from a model of such simplicity.

We turn now to the issue of correlations. If the grain orientations were statistically independent, i.e., uncorrelated, one would have the simple result $\langle h_n^2 \rangle = n\varepsilon$, while (5) implies that $\langle h_n^2 \rangle$ grows much more slowly than n. The orientational displacements of each grain are thus *fully anticorrelated.* Fig. 2 shows that the orientation correlations $c_{m,n} = \langle \sigma_m \sigma_n \rangle$ scale as [1,2]

$$c_{m,n} \approx \delta_{m,n} - \frac{1}{W_m W_n} F\left(\frac{n-m}{W_m W_n}\right), \tag{6}$$

where the function F is such that $\int_{-\infty}^{+\infty} F(x)\,\mathrm{d}x = 1$. The fluctuations of the orientational displacements are therefore asymptotically *totally screened:* $\sum_{n \neq m} c_{m,n} \approx -c_{m,m} = -1$.

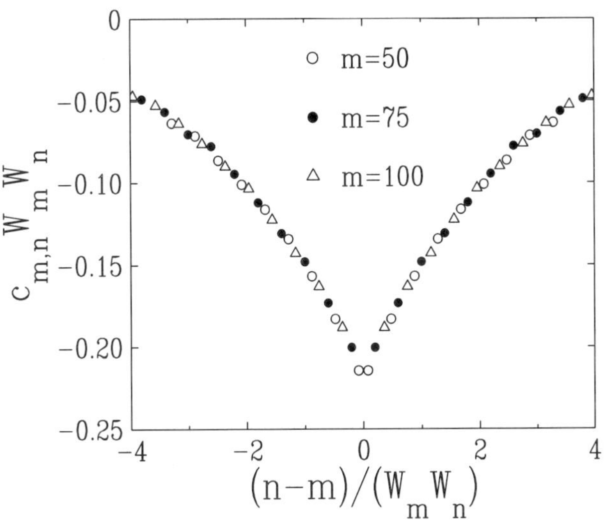

Figure 2. Scaling plot of the orientation correlation function $c_{m,n}$ for $n \neq m$ in the zero-temperature steady state with $\varepsilon = 1$, demonstrating the validity of (6) and showing a plot of (minus) the scaling function F (after [1,2]).

Summing up, we see that the (orientational) displacements of regular grains are anticorrelated[4] within a *dynamical cluster* [13] whose size scales as $n^{2/3}$. Grains separated by

[4]Correspondingly, from a kinetic viewpoint, these results may be interpreted in terms of the time $n^{2/3}$ spent by a walker bouncing back and forth between the walls of a cage, where his steps are consequently anticorrelated one with the other.

greater than the cluster radius are orientationally screened from each other, i.e., the screening length also goes as $n^{2/3}$. Consistently, the order parameter $Q_n = \langle \sigma_n \operatorname{sign} h_n \rangle$, proportional to the ratio of the screening length to the total length, goes as $n^{2/3}/n \sim n^{-1/3}$.[5]

Similar anticorrelations[6] in grain displacements have been observed in hard-sphere simulations of shaken powders close to jamming [13]; the observation was that grain displacements *along* the direction of vibration were strongly anticorrelated, while transverse to it, they were uncorrelated. In the absence of free voids, compaction in the simulations happened by the phenomenon of *bridge collapse*; the vibrations coupled to the longitudinal displacements of the grains, allowing upper and lower grains in a bridge to collapse onto each other, thus minimising the trapped void space between them. This led to the observed longitudinal anticorrelations; the low intensities of vibration on a jammed granular bed did not allow for a coupling with transverse granular displacements. Given the close agreement of these earlier results [13] with our present ones, we can in hindsight justify our choice of a columnar model to model granular compaction in the jamming limit.

Finally, we point out that the landscape of visited configurations in the steady state of density fluctuations has a fractal-like structure [1,2]. Fig. 3 shows that, on whichever scale we look, some configurations are clearly visited far more often than others. It turns out that the most visited configurations are the ground states of the system (empty circles). We suggest that this behaviour is generic: i.e., *the dynamics of compaction in the jammed state leads to a microscopic sampling of configuration space which is highly non-uniform, so that the ground states are visited most frequently.* This might be expected when the system is constrained to compact. However, it should be noted that under continuous shaking, the system cannot rest in a ground state even when one is found; it evolves constantly, and thereby generates the observed density fluctuations. It should also be noted that the entropy reduction $\Delta S \sim n^{1/3}$ resulting from this fine structure is subextensive, and therefore negligible with respect to the free entropy $S_{\text{flat}} = n \ln 2$. Our model thus provides a natural reconciliation between, on the one hand, the intuitive perception that not all configurations can be equally visited during compaction in the jamming limit; and, on the other, the flatness hypothesis of Edwards, which states that for large enough systems, the entropic landscape of visited configurations is flat [15].

4. Low-temperature dynamics

We now turn to the investigation of the low-temperature dynamics of the model. Our main finding is the observation of *intermittency* in the position of the boundary, or surface, layer; this has recently been observed in experiments of vibrated granular beds [16].

For rational ε, the presence of a finite but low shaking intensity does not change much; the zero-temperature dynamics is in any case stochastic, and low-temperature dynamics merely increases the effect of noise. However, for irrational ε, low-temperature dynamics introduces an intermittency in the position of a surface layer, which separates a quasiperi-

[5] By contrast, when ε is irrational, earlier orientations influence *all* successive ones, as the orientation correlations c_{mn} do not decay to zero. The order parameter is $Q_n = 1$ identically while, loosely speaking, the screening length scales as n.

[6] Treating n as time in a random walk, these anticorrelations also recall the temporal anticorrelations observed in recent experiments investigating cage properties near the colloidal glass transition (see e.g. [14]).

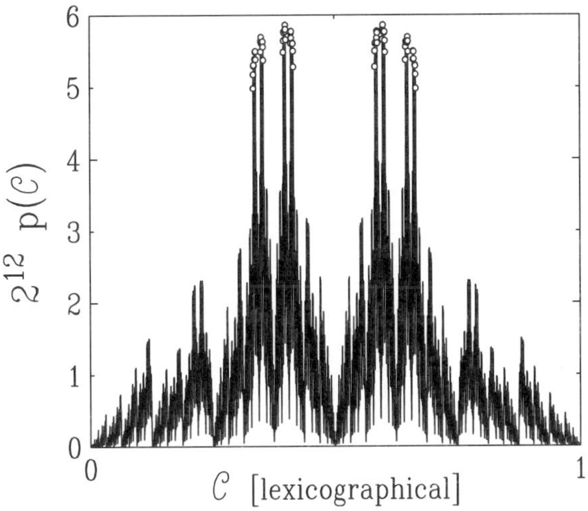

Figure 3. Plot of the normalised probabilities $2^{12}\,p(\mathcal{C})$ of the configurations of a column of 12 grains in the zero-temperature steady state with $\varepsilon = 1$, against the configurations $\mathcal{C}$ in lexicographical order. The empty circles mark the $2^6 = 64$ ground-state configurations, which turn out to be the most probable (after [1,2]).

odically ordered region near the surface from a steady state of density fluctuations in the bulk.

This happens as follows: when the shaking energy Γ is such that it does not distinguish between a very small void h_n and the strict absence of one, the site n 'looks like' a point of perfect packing. The grain at depth n then has the freedom to point the 'wrong' way; we call such sites *excitations*, using the thermal analogy. The probability of observing an excitation at site n scales as $\Pi(n) \approx \exp(-2|h_n|/\Gamma)$; the sites n so that $|h_n| \sim \Gamma \ll 1$ will be preferred and thus dominate the low-temperature dynamics. These preferred sites are such that $n\Omega$ is closest to an integer, making them look most like points of perfect packing; misalignment vis-a-vis (3) thus costs the least. The uppermost excitation is propagated ballistically (cf. zero-temperature (irrational ε) dynamics) until another excitation is nucleated above it; its instantaneous position $\mathcal{N}(t)$ denotes the layer at which shape effects are lost in thermal noise, i.e., it separates an upper region of quasiperiodic ordering from a lower region of density fluctuations (5).

Fig. 4 shows a typical sawtooth plot of the instantaneous depth $\mathcal{N}(t)$, for a temperature $\Gamma = 0.003$. The *ordering length*, defined as $\langle\mathcal{N}\rangle$, is expected to diverge at low temperature, as excitations become more and more rare; we find in fact [1,2] a divergence of the ordering length at low temperature of the form $\langle\mathcal{N}\rangle \sim 1/(\Gamma|\ln\Gamma|)$. This length is a kind of finite-temperature equivalent of the 'zero-temperature' length $\xi_{\rm dyn}$, as it divides an ordered boundary layer from a lower (bulk) disordered region. Within each of these boundary layers, the relaxation is *fast*, and based on single-particle relaxation, i.e., individual particles attaining their positions of optimal local packing [7,9]. The *slow* dynamics of cooperative relaxation only sets in for lengths *beyond* these, when the lengths over which packing needs to be optimised become non-local.

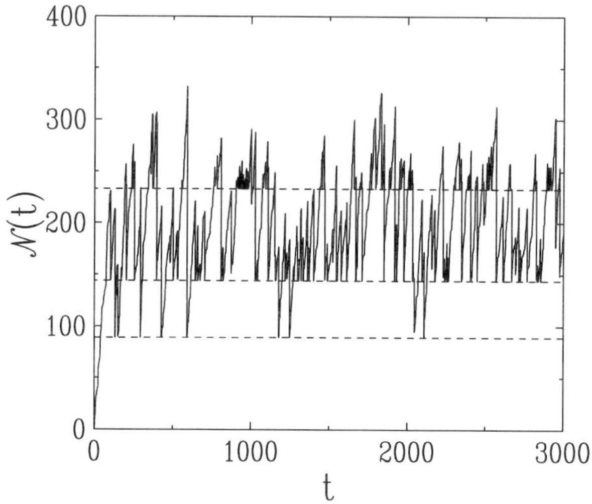

Figure 4. Plot of the instantaneous depth $\mathcal{N}(t)$ of the ordered layer, for $\varepsilon = \Phi$ (the golden mean) and $\Gamma = 0.003$. Dashed lines: leading nucleation sites given by Fibonacci numbers (bottom to top: $F_{11} = 89$, $F_{12} = 144$, $F_{13} = 233$) (after [1,2]).

5. Discussion

In this work, we have tried to explain why jagged grains (such as would be found in gravel) pack better than smooth ones (such as bricks), when submitted to gentle shaking. Our control parameter is ε, the void space wasted by the disordered orientation of a grain; we have shown that irrational and rational values of this parameter lead to very different effects.

The ground state of jagged grains (irrational ε) is unique and only quasiperiodically ordered; that for smooth grains (rational ε) is highly degenerate and perfectly ordered. This is intuitively obvious; the many rough edges of irregular grains need very special orientations to click into a perfect packing, whereas rectangular bricks, for example, can be arranged in many ways so that they are perfectly packed.

The very perfection of the ground states for regularly shaped grains makes them impossible to retrieve stochastically, even in the limit of zero-temperature dynamics; instead, the effect of the latter is to give rise to density fluctuations, as predicted by our model, and observed experimentally [12]. The 'rough-and-ready' nature of the ground state of jagged grains is by contrast quickly (ballistically) retrievable. Clearly, a sharp distinction between neighbouring rational and irrational values of ε only makes sense for an infinitely deep system; for a finite column made of N grains, the distinction is rounded off by finite-size effects. In particular, the characteristic features of any 'large' rational ε are no longer observed when the period $p+q$ becomes larger than N.

The density fluctuations seen in the case of regular grains have a slightly non-Gaussian nature [8] caused by their (anti)correlations; this is reminiscent of dynamical heterogeneities in strongly compacted granular media [13], as well as temporal anticorrelations in cages [14]. Also, while the macroscopic entropy [17] of visited configurations in the steady state of density fluctuations is consistent with Edwards' 'flatness' hypothesis [15], the microscopic configurational landscape is very rugged, with the most visited configurations corresponding to the ground states - as might be expected for compaction in the jamming limit. Lastly, the low-temperature dynamics for irrational ε leads to an intermittency of the boundary layer separating quasiperiodic order from disordered density fluctuations; for shaking at sufficiently low intensities, it should be possible to test our detailed predictions [1,2] for intermittency using irregularly shaped grains.

Remarkably, many of the above features were obtained at a *qualitative* level in the glassy regime of a much simpler model [7]. On the one hand, this allows us to speculate that the shape-dependent ageing phenomena seen there could be retrieved here, i.e., that conventional ageing phenomena would only be seen for irregular grains (irrational ε). On the other hand, it is tempting to ask if the *directional causality* of the dynamical interactions present in this model and the earlier one [7], could be responsible for their qualitative similarity, and thus be a necessary ingredient for modelling 'glassiness'?

Acknowledgements

AM warmly acknowledges the hospitality of the Service de Physique Théorique, Saclay, where most of this work was conceived.

REFERENCES

1. A. Mehta and J.M. Luck, J. Phys. A **36**, L365 (2003).
2. J.M. Luck and A. Mehta, Eur. Phys. J. B **35**, 399 (2003).
3. M. Mézard, G. Parisi, and M.A. Virasoro, *Spin Glass Theory and Beyond* (World Scientific, Singapore, 1987).
4. S.F. Edwards, in *Granular Matter: An Interdisciplinary Approach*, ed. A. Mehta (Springer, New York, 1994).
5. P.G. de Gennes, Rev. Mod. Phys. **71**, S374 (1999).
6. E. Marinari, G. Parisi, F. Ricci-Tersenghi, and F. Zuliani, J. Phys. A **34**, 383 (2001); M. Mézard, Physica A **306**, 25 (2002); G. Biroli and M. Mézard, Phys. Rev. Lett. **88**, 025501 (2002); A. Lawlor, D. Reagan, G.D. McCullagh, P. De Gregorio, P. Tartaglia, and K.A. Dawson, Phys. Rev. Lett. **89**, 245503 (2002).
7. P.F. Stadler, J.M. Luck, and A. Mehta, Europhys. Lett. **57**, 46 (2002); P.F. Stadler, A. Mehta, and J.M. Luck, Adv. Complex Systems **4**, 429 (2001).
8. E.R. Nowak, A. Grushin, A.C.B. Barnum, and M.B. Weissman, Phys. Rev. E **63**, 020301 (2001).

9. J. Berg and A. Mehta, Europhys. Lett. **56**, 784 (2001); Phys. Rev. E **65**, 031305 (2002).
10. J.D. Bernal, Proc. R. Soc. London A **280**, 299 (1964).
11. R.L. Brown and J.C. Richards, *Principles of Powder Mechanics* (Pergamon, Oxford, 1970).
12. E.R. Nowak, J.B. Knight, M. Povinelli, H.M. Jaeger, and S.R. Nagel, Powder Technology **94**, 79 (1997); E.R. Nowak, J.B. Knight, E. Ben-Naim, H.M. Jaeger, and S.R. Nagel, Phys. Rev. E **57**, 1971 (1998).
13. A. Mehta and G.C. Barker, Phys. Rev. Lett. **67**, 394 (1991); G.C. Barker and A. Mehta, Phys. Rev. A **45**, 3435 (1992); Phys. Rev. E **47**, 184 (1993); A. Mehta and G.C. Barker, Europhys. Lett. **27**, 501 (1994); J. Phys. Cond. Matt. **12**, 6619 (2000).
14. E.R. Weeks and D.A. Weitz, Chem. Phys. **284**, 361 (2002).
15. S.F. Edwards, in *Challenges in Granular Physics*, eds. T.C. Halsey and A. Mehta (World Scientific, Singapore, 2003).
16. G. Caballero et al, this volume.
17. R. Monasson and O. Pouliquen, Physica A **236**, 395 (1997).

Unifying Concepts in Granular Media and Glasses
A. Coniglio, A. Fierro, H.J. Herrmann and M. Nicodemi (editors)

On pre-asymptotic aging in finite dimensional spin glasses.

S. Franz [a] V. Lecomte [b] R. Mulet [c]

[a] International Centre for Theoretical Physics, Strada Costiera 11, P.O. Box 586, I-34100 Trieste, Italy

[b] Département de Physique, École normale supérieure de Cachan, avenue du Président Wilson, 94230 Cachan, France

[c] "Henri Poincare" Chair of Complex Systems and Superconductivity Laboratory, Physics Faculty-IMRE, University of Havana, La Habana, Cuba

We study the off-equilibrium dynamics in the pre-asymptotic aging regime of the two, three and four dimensional Edwards-Anderson spin glass model. We define an exotic correlation function and the corresponding response, and we compare the resulting fluctuation-dissipation ratio (FDR) with the usual one obtained by the ordinary spin-spin correlation function and relative response. In all systems we find that that after a short transient the two corresponding fluctuation-dissipation ratios coincide at equal times. Our findings support the idea that the FDR defines a unique set of effective temperatures for degrees of freedom evolving on the same time scale. In addition we show that in 2D, as it happens for the usual FDR, the new dynamic FDR at finite time can be relates to the static overlap probability function (OPF) of systems of suitable finite size.

1. Introduction

Glasses are trapped for long times in off-equilibrium states. Their relaxation assumes the character of aging, becoming slower and slower as the system evolves towards more and more optimized regions of the configuration space. A fundamental open question is to give a statistical description of aging, understanding how the configuration space is sampled during this process[1–3,6].

In connection with this problem, in recent time a lot of attention has been devoted to the study of the violations of the fluctuation dissipation theorem. Given the correlation function $C(t,t_w)$ of a certain observable A, and its conjugated response function $\chi(t,t_w)$ describing the effect at time t of a field conjugated to A acting from time 0 to time t_w, one defines the fluctuation-dissipation ratio (FDR) $X(t,t_w)$ from the relation:

$$X(t,t_w) = T\frac{\partial C(t,t_w)/\partial t_w}{\partial \chi(t,t_w)/\partial t_w}. \tag{1}$$

Deviation of the FDR from unity indicate off-equilibrium behaviour. In generic off-equilibrium situations one can expect the FDR to depend on the observable quantity A at hand.

The theory of mean field spin glasses, has shown the existence of systems where the FDR admits a non-trivial long time limit, depending on the path by which t and t_w are sent to infinity. If t and t_w are sent to infinity fixing the values of the correlation $C(t, t_w)$ one gets a function

$$x(q) = \lim_{t,t_w \to \infty} X(t, t_w)|_{C(t,t_w)=q}. \tag{2}$$

Importantly, for these systems the FDR for different couples of conjugated correlations and responses admit the same limit along the same path. If we consider a correlation function $C_A(t, t_w)$ corresponding to an observable A and a correlation function $C_B(t, t_w)$ corresponding to an observable B, asymptotically the FDRs $X_A(t, t_w)$ and $X_B(t, t_w)$ tend to coincide. More precisely, if we define an auxiliary limiting function $q_B(q_A) = \lim_{t,t_w \to \infty} C_B(t, t_w)|_{C_A(t,t_w)=q_A}$, the following relation holds

$$x_B(q_B(q_A)) = x_A(q_A). \tag{3}$$

At present it is not clear if in some finite dimensional system, with non singular interactions, the FDR admits a non-trivial limit. If this happens however, this has to be related to static ergodicity breaking. In ref. [4] it has been shown that $x(q)$ can be related to the overlap probability function (OPF) $P(q)$ [5] describing the statistics of the pure states at equilibrium not related by a symmetry of the Hamiltonian:

$$\frac{dx(q)}{dq} = P(q). \tag{4}$$

This relation, which holds trivially in ergodic systems where both $x(q)$ and $P(q)$ are equal to δ-functions are both either non-trivial or trivial could assume a non trivial content in systems with aging persisting for all times.

The identity (4) and eq. (3), support the interpretation of the ratios $T/x(q)$ for different values of q as effective temperatures governing the heat exchanges between slowly evolving modes[7].

Despite the attractiveness of this description, its strict validity is confined to the asymptotic regime where the energy and all other single time observables are close to their equilibrium values. However, both in simulations and in experiments one is often very far from such an ideal long time regime.

Extrapolations of numerical simulations of 3 and 4 dimensional spin glasses comparing the OPF of small systems and the FDR for finite times indicate the non-triviality -and consistently the identity- of both functions [8]. However these extrapolations have been criticized in a series of papers explicitly displaying systems where the OPF is trivial in the thermodynamic limit, but resembles the one of mean field for small enough volumes [9]. Moreover, in the experimental side it is manifest that many aging systems are found in pre-asymptotic regimes. For example the experiments of Ocio and Herisson [10] show FD curves that strongly depend on the waiting time and very far from a reasonably guessed asymptote. In structural glasses where the observed values of one time quantities are cooling rate dependent, the situation is even worse.

It is therefore of great interest to inquire if the concepts valid for the asymptotic regime can be adapted to get an adequate picture of the dynamics on much shorter time scales.

It is possible to hypothesize that the identity between static and dynamic FDR found in 3 and 4 dimension could be the effect of a slowly growing length $\xi(t_w)$ over which the system has effectively equilibrated. $X(q, t_w)$ would then approximately respect the relation (4) with the OPF $P(q, L = \xi(t_w))$ of a system of size $L = \xi(t_w)$. This extension put forward by A. Barrat and L. Berthier in [11] would suggest the approximate validity at finite time of a quasi-equilibrium picture of the aging dynamics in which quasi-states with equal free-energy are selected with equal probability, and the static-dynamic equivalence would just reflect the properties of the equilibrium landscape of finite size system. In [11] and [12] it was found that indeed the conjectured relation among finite time FDR and finite size OPF quite remarkably exists.

At a first sight these findings appear in contrast with the relation among effective temperatures and time scale separation. For finite times, strict time scale separation is not possible and slow modes have to exchange heat in order to eventually equilibrate. However the picture can be saved hypothesizing an "adiabatic cascade", in which modes evolving at the same rate appear able to equilibrate at their effective temperature before exchanging heat with neighbouring modes. In such conditions FDR corresponding to different quantities should approximately coincide. In a recent paper [13] we have tested this hypothesis in the the 2D Edwards-Anderson (2DEA) model where as mentioned aging is interrupted after a finite relaxation time. Here we summarize this work, and we present additional data on the early time pre-asymptotic regime of the Edwards-Anderson model in three and four dimension.

The remaining of the paper is organized as follows. In the next section, we introduce the relevant quantities, then we present and discuss the results of the simulations and finally the conclusions are outlined.

2. Definition of the observables.

In spin systems, in usual FD plots, one compare the spin-spin autocorrelation function $C(t, t_w) = N^{-1} \sum_i \langle S_i(t) S_i(t_w)\rangle$ to the response to an external field: the "zero field cooled" susceptibility to small local Gaussian fields h_i with variance h_o^2, introduced in the systems at time t_w $\chi(t, t_w) = \frac{1}{N h_o^2} \sum_i \langle h_i S_i(t)\rangle$.

The task of this section is to define a different couple of correlation/response, non trivially related to the former. Consider two copies of Edwards-Anderson spin glass systems, with identical number of spins, and identically distributed, but independent quenched disorder and coupled by a random field R_i. The Hamiltonian is then

$$H = \sum_{i,j} J^1_{<ij>} S^1_i S^1_j + \sum_{<i,j>} J^2_{ij} S^2_i S^2_j + \sum_i R_i S^1_i S^2_i \tag{5}$$

where the spins are Ising variables, the quenched disorder J^1_{ij} and J^2_{ij} in copies 1 and 2 are i.i.d. centered Gaussian variables with unit variance, and the variables R_i which couple spins with identical label in the two copies are chosen randomly with values $R_i = \pm K$.

We now define the cross-correlation function,

$$C_{cross}(t, t_w) = (2N)^{-1} \sum_i \overline{\langle (S^1_i(t) S^2_i(t_w) + S^2_i(t) S^1_i(t_w)) R_i\rangle} \tag{6}$$

and its conjugated cross-response

$$\chi_{cross}(t,t_w) = \frac{1}{2Nh_o^2K^2}\sum_i \overline{\langle (h_i^2 S_i^1(t) + h_i^1 S_i^2(t)) R_i \rangle} \quad (7)$$

Where the $\langle \ldots \rangle$ indicates an average over the initial conditions and $\overline{\cdots}$ over the disorder. As opposed to the cross-correlation and response, the usual correlation and response will be called "direct".

The justification for using the cross functions can be found in their linear response valid for small values of K:

$$\begin{aligned} C_{cross}(t,t_w) &= K^2\beta[\int_0^{t_w} ds C(t,s)R(t_w,s) \\ &\quad + \int_0^t ds\, C(t_w,s)R(t,s)\,] \\ R_{cross}(t,t_w) &= K^2\beta\int_{t_w}^t ds\, R(t,s)R(s,t_w), \end{aligned} \quad (8)$$

showing that there is not a trivial relation between these quantities and the direct ones valid independently of the dynamical process.

In comparing static OPF and dynamic FDR for the cross-functions, an additional grain of salt is needed. One can notice that while for all times $C(t,t) = 1$, the value of $C_{cross}(t,t)$ will depend on time. In such conditions a better suited quantity to use in FD plots instead of the correlation is the "difference"-function $B_{cross}(t,t_w) = \frac{1}{2}[C_{cross}(t,t) + C_{cross}(t_w,t_w) - 2C_{cross}(t,t_w)]$. The static analogous of $B(t,t_w)$ to be used as variable in the OPF can be defined, given two configurations $\mathbf{S}$ and $\mathbf{S}'$, by the quantity $b_{cross}(\mathbf{S},\mathbf{S}') = \frac{1}{2}[q_{cross}(\mathbf{S},\mathbf{S}) + q_{cross}(\mathbf{S}',\mathbf{S}') - 2q_{cross}(\mathbf{S},\mathbf{S}')]$.

3. Results and Discussion

We compared the behaviour of the cross and direct quantities in the Edward Anderson model, in 2, 3 and four dimensions choosing the value $K = 1/2$ as the strength of the coupling term. For this large value of K we are out of the linear response regime that allowed us to derive the explicit form of the cross-quantities as function of the usual ones, but even if the relations (8) does not hold, there is no reason to believe that the relation between X_{cross} and X becomes trivial.

We first considered the 2D model, and compared the dynamic FD plot obtained at different times with the static one for small volumes obtained integrating twice the OPF. The dynamic results were obtained through regular metropolis simulations of a 512×512 systems where finite size effects where not seen on the time scales we probed. The static data where obtained using parallel tempering on small samples. The results for $T = 0.43$ are summarized in figure 1, where we trace the FD characteristics both for the direct and the cross functions. The direct characteristics is completely analogous to the one found by Berthier and Barrat showing that tuning appropriately the waiting times one can approximately superimpose the aging characteristic with the equilibrium ones for finite volume. The cross characteristics, remarkably, shows that the agreement of the cross functions is perfectly comparable with the one of the direct one. In our view this fact supports the idea of a characteristic length $\xi(t_w)$, growing with the waiting time, over which the system is equilibrated, and that the equilibrium is similar to that of an isolated system of that size. If that correspond to the truth both direct and cross functions should define the same effective temperature. We then tested the idea that $X(t,t_w)$ and $X_{cross}(t,t_w)$ could have the same behaviour for finite times. In order to do that we needed very good data for correlations and especially responses, in order to extract derivatives from the FD plots. This has been achieved averaging each curve over 190 different samples. The results in 2D can bee seen in figure 2. For very short waiting times $t_w = 10$ one sees that the curves are different, while, starting fro waiting times as short as $t_w = 100$ both FDR coincide within numerical error.

We tested the same property in three and four dimension for short times t_w, and large systems so that finite size effects were negligible. In figures 3 and 4 we plot the FD characteristic for both models. It can be seen that the curves are strongly dependent of t_w. Despite this fact, we can see in fig. 5 and 5 that the slopes of the curves coincide at equal times. This indicates that in this regime while both FDR are time dependent, degrees of freedom that evolve on the same time scales are in mutual equilibrium.

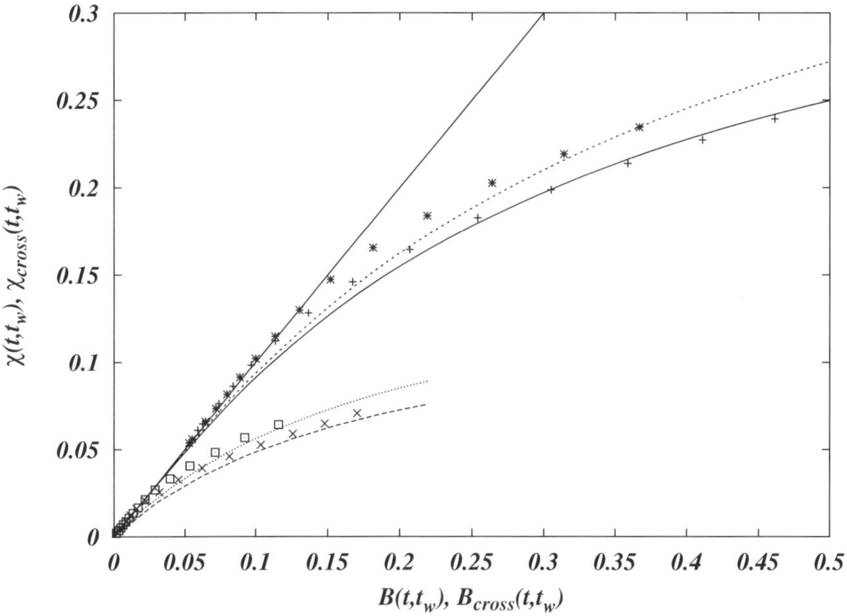

Figure 1. Fluctuation dissipation plot for the 2DEA model using B and B_{cross} as abscissas. The agreement of the dynamical and the static data for the cross quantity is comparable to the direct ones.

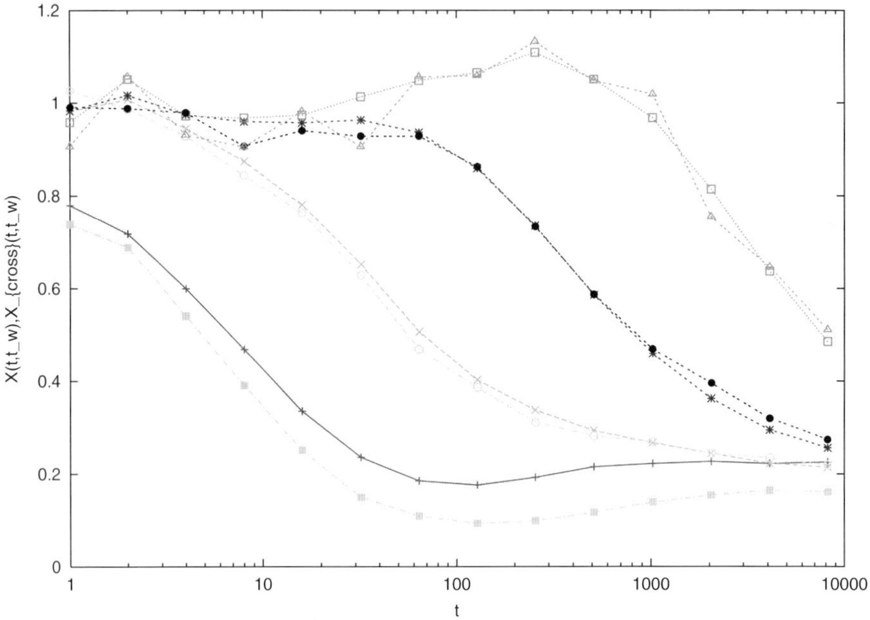

Figure 2. Comparison between the FDR for the cross and direct quantities in the 2D EA model. The temperature is $T = 0.43$ and the waiting times are, from bottom to top $t_w = 10, 100, 1000, 10000$. Notice that while the two FDRs at $t_w = 10$ are markedly different, both quantity coincide for $t_w \geq 100$.

4. Conclusions

In this paper we review our results on the analysus of cross-correlation and response functions in the pre-asymptotic aging regime of the 2D Edwards-Anderson spin glass model. We show how the behaviour of these function supports the idea that even at finite waiting times there is a set of uniquely defined effective temperatures, and the idea of a growing length over which the system has reached local equilibrium. In addition we presented new data in 3D and 4D on pre-asymptotic aging, finding that also in that case direct and cross quantity are in mutual equilibrium with each other. These findings suggest that the ideas coming from the asymptotic analysis of mean field systems, performed in the thermodynamic limit for infinite times, could be useful to understand the behaviour of the glassy dynamics of finite dimensional systems at finite times.

Acknowledgments

We thanks A. Barrat and L. Berthier for useful discussions and suggestions. This work was supported in part by the European Community's Human Potential programme under

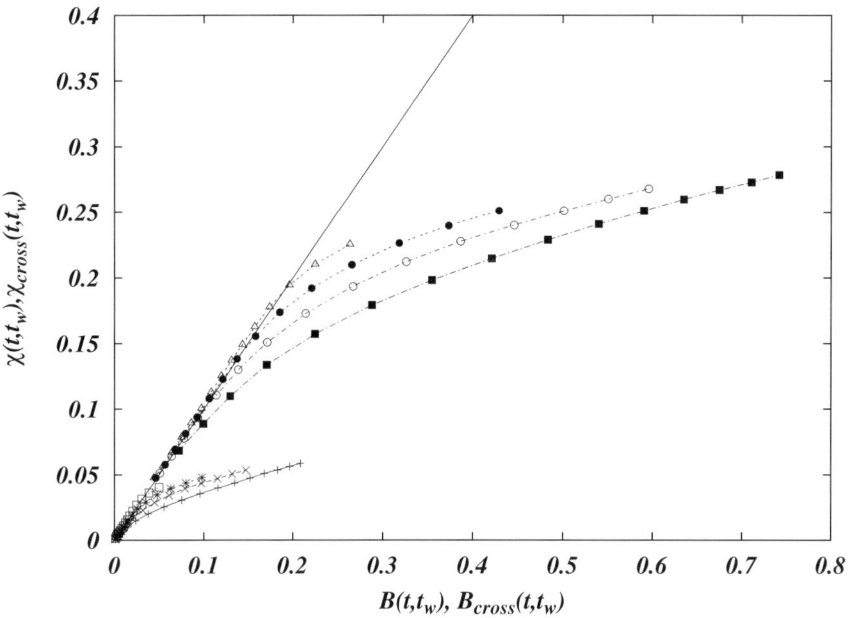

Figure 3. Fluctuation dissipation plot for the 3D EA model using B and B_{cross} as abscissas. The temperature is $T = .43$ and the waiting times are $t_w = 10, 100, 1000, 10000$. We see that on that time scales the curves are strongly t_w-dependent.

contract "HPRN-CT-2002-00319 STIPCO".

REFERENCES

1. S. Franz and M.A. Virasoro, J. Phys A: Math. Gen. **33** (2000) 891
2. G. Biroli and J. Kurchan Phys. Rev. E **64** 016101 (2001)
3. A. Barrat, J. Kurchan, V. Loreto and M. Sellitto Phys. Rev. Lett. **85** (1999) 5034; Phys. Rev. E **63** 051301 (2001).
4. S. Franz, M. Mezard, G. Parisi and L. Peliti, Phys. Rev. Lett. **81** (1998) 1758, J. Stat. Phys. **97**,459 (1999)
5. M. Mezard, G. Parisi and M. A. Virasoro, *Spin Glass Theory and Beyond*, Singapore: World Scientific (1987)
6. S. Franz "How glasses explore the configuration space" Proceedings of the "3rd workshop on non-equilibrium phenomena in supercooled fluids, glasses and amorphous materials" 22-27 September, Pisa. Preprint cond-mat/0212091
7. L. Cugliandolo, J. Kurchan and L. Peliti, Phys. Rev. E **55** 3898 (1997).
8. E. Marinari, G. Parisi, F. Ricci-Tersenghi, J. J. Ruiz-Lorenzo (1998) J. Phys. A **31**,

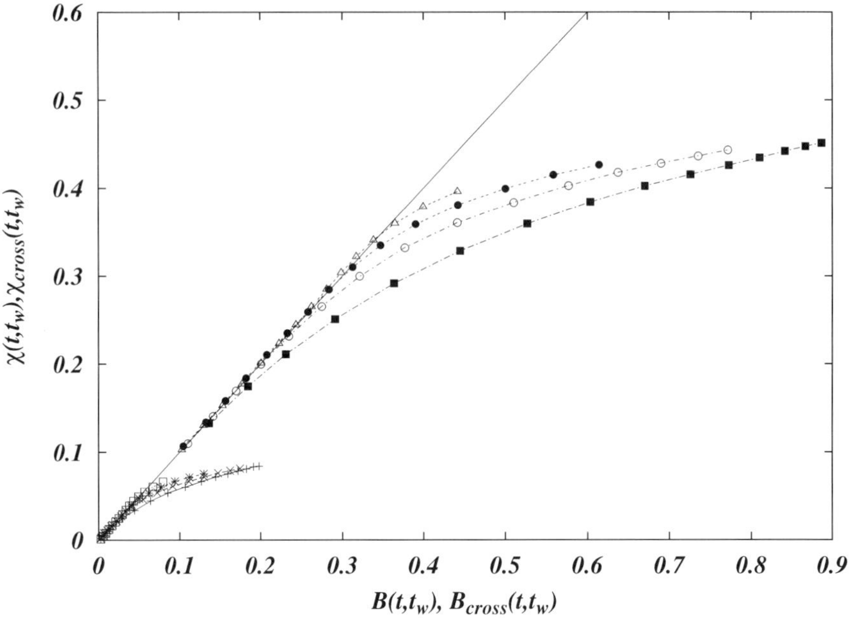

Figure 4. Fluctuation dissipation plot for the 4D EA as in fig. 3 . The temperature is $T = .43$ and the waiting times are $t_w = 10, 100, 1000, 10000$. Also in this case we see a strong dependence of the curves on t_w.

2611
9. M. Moore, H. Bokil and B. Drossel Phys. Rev. Lett. **81** 4252 (1998)
10. D. Herisson D and M. Ocio, Phys. Rev. Lett. **88** 25720 (2002)
11. L. Berthier and A. Barrat, Phys. Rev. Lett. **89**, 095702 (2002)
12. L. Berthier, J. Phys. A **36**, 10667 (2003)
13. S. Franz, V. Lecomte and R. Mulet, to appear in PRE, preprint cond-mat/0309194.

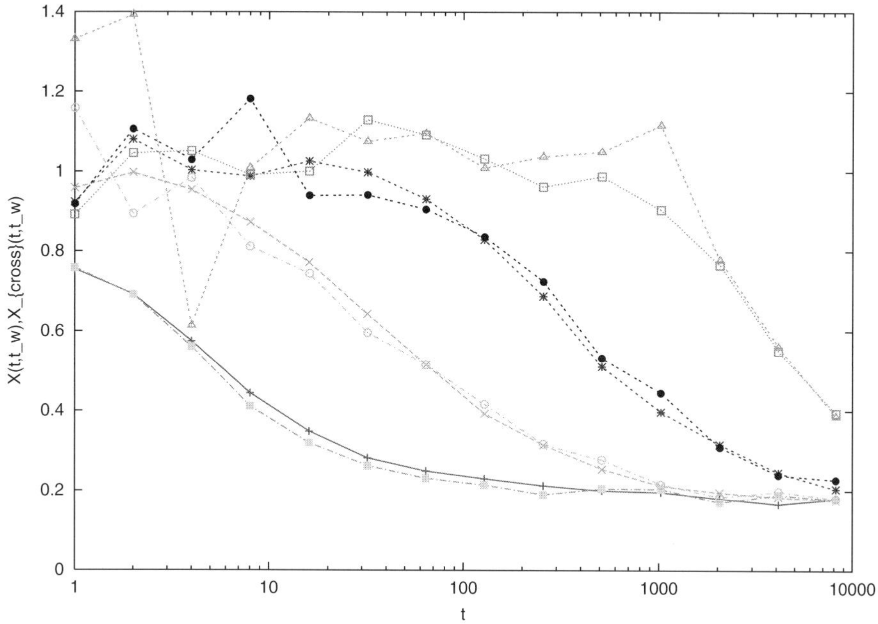

Figure 5. Same as fig. 2 in three dimension. The temperature is $T = 0.43$ and the waiting times are, from bottom to top $t_w = 10, 100, 1000, 10000$. In this case the curves coincide even for the shortest t_w. The quality of the data at longer waiting times is poorer due to less statistics.

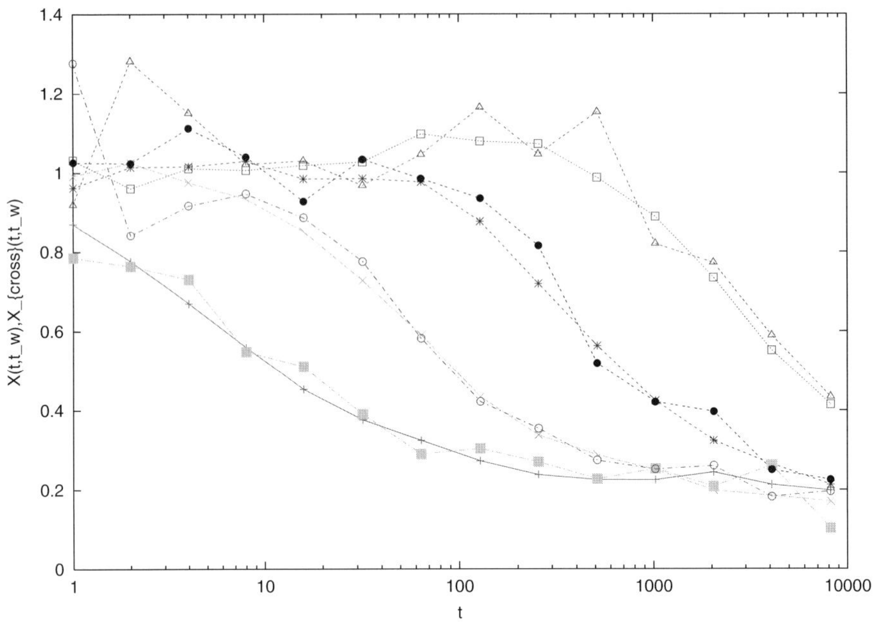

Figure 6. Direct comparison of the FDR for the cross and direct quantities in the 4D EA model. The temperature is $T = 0.43$ and the waiting times are, from bottom to top $t_w = 100, 1000, 10000$. After a shor transient both quantities do coincide.

Unifying Concepts in Granular Media and Glasses
A. Coniglio, A. Fierro, H.J. Herrmann and M. Nicodemi (editors)

Stimulated and spontaneous relaxation in glassy systems

F. Ritort
Faculty of Physics, University of Barcelona, Diagonal 647, 08028 Barcelona, Spain and Department of Physics, University of California, Berkeley CA 94720, USA
E-Mail:ritort@ffn.ub.es

Recent numerical simulations of a disordered system [5] have shown the existence of two different relaxational processes (called stimulated and spontaneous) characterizing the relaxation observed in structural glasses. The existence of these two processes has been claimed to be at the roots of the intermittency phenomenon observed in recent experiments. Here we consider a generic system put in contact with a bath at temperature T and characterized by an adiabatic slow relaxation (i.e. by a negligible net heat flow from the system to the bath) in the aging state. We focus on a simplified scenario (termed as partial equilibration) characterized by the fact that $T = 0$ (where only the spontaneous process is observable) and whose microscopic stochastic dynamics is ergodic when constrained to the constant energy surface. Three different effective temperatures can be defined: a) from the fluctuation-dissipation theorem (FDT), $T_{\rm eff}^{\rm FDR}$, b) from a fluctuation theorem describing the statistical distribution of heat exchange events between system and bath, $T_{\rm eff}^{\rm FT}$ and c) from a set of observable-dependent microcanonical relations in the aging state, $T_{\rm eff}^{\rm MR}$. In a partial equilibration scenario we show how all three temperatures coincide reinforcing the idea that a statistical (rather than thermometric) definition of a non-equilibrium temperature is physically meaningful in aging systems. These results are explicitly checked in a simple model system.

1. Introduction

Efforts to extend well established concepts in equilibrium thermodynamics to the non-equilibrium domain have repeatedly appeared many times in the past in different contexts. An idea that has attracted the attention of physicists for quite a long time is the concept of a temperature applied to non-equilibrium states [1]. In equilibrium, the notion of temperature can be covered from two different perspectives. On the one hand, there is the thermodynamic approach where temperature is defined as a parameter that characterizes classes of systems in mutual thermal equilibrium. The usefulness of this *thermometric* definition relies on the validity of the zeroth law of thermodynamics. On the other hand, there is a statistical approach (ensemble theory) where temperature can be defined from the properties of individual systems without any reference to the mutual equilibrium property. The *statistical* temperature is defined as the inverse of the energy gradient of the configurational entropy measured over a constant energy surface. The maximum entropy postulate relates the statistical notion of the temperature to the thermometric one. The statistical and thermodynamic concepts look equivalent but they are not as the latter requires a specific behavior of the different systems when put in mutual contact. The thermodynamic definition of a temperature represents a stronger condition than the statistical one.

An interesting category of systems that has recently received considerable attention are glassy systems in their aging regime. The aging regime is a slow relaxational process characterized by the extremely small net amount of heat delivered by the system to the bath in contact. It has been suggested [2] that in aging systems a thermometric definition of non-equilibrium temperature is meaningful for a thermometer responding to low-enough frequencies. We suspect that this definition is too strong and might be wise to investigate a *low-level* statistical definition of a non-equilibrium temperature.

The purpose of this paper is to show that a statistical definition of non-equilibrium or effective temperature (rather than thermometric) can be rescued for a particular category of aging systems characterized by adiabatic slow relaxation. In this category of systems relaxation is imposed to be ergodic when constrained to a given energy shell. At $T = 0$ this condition defines what we term as *partial equilibration* scenario. For this class of systems three different definitions of an effective temperature are possible: a) from the fluctuation-dissipation theorem (FDT), $T_{\rm eff}^{\rm FDR}$; b) from the fluctuation theorem applied to the statistical distribution of heat exchange events between system and bath, $T_{\rm eff}^{\rm FT}$; c) from a set of microcanonical relations in the aging state, $T_{\rm eff}^{\rm MR}$. In a partial equilibration scenario we show how all these temperatures coincide reinforcing the idea that a statistical definition of a non-equilibrium temperature is physically meaningful in aging systems. We check all statements and results in a simple model of glassy system where the partial equilibration scenario holds and explicit computations can be done.

2. Adiabatic relaxation

Let us consider a system that is prepared in a non-equilibrium state by placing it in contact with a thermal bath at low temperature (quenching protocol). The system will relax and release heat to the bath in a process that can span from several minutes to millions of years. Only when the system has reached equilibrium the net heat current from the system to the bath vanishes. All along the paper we will refer to this relaxational regime as the aging regime and the corresponding non-equilibrium state as the aging state. The time after the quench will be referred as the age of the system and will be denoted by one or two of the variables s and t depending on whether one-time or two-time quantities quantities are considered. We adopt the convention $t > s$.

Several definitions and quantities seem appropriate to put the discussion in perspective [1]. Let us consider a system described by an energy function $E(\mathcal{C})$ where $\mathcal{C}$ denotes a generic configuration. The system is in contact with a thermal bath and the microscopic stochastic dynamics is both ergodic and satisfies the detailed balance property. $P_t(\mathcal{C})$ will denote the probability for the system to be in the configuration $\mathcal{C}$ at time t. The average value of an observable A at time t will be denoted by $A(t)$ and is given by,

$$A(t) = \sum_{\mathcal{C}} A(\mathcal{C}) P_t(\mathcal{C}) \quad . \tag{1}$$

[1] For a detailed presentation of several of these concepts see [3]

Often the time argument in $A(t)$ will be dropped off and we will simply write A, with the clear understanding that in general it is a time-varying quantity. $P(\mathcal{C}',t|\mathcal{C},s)$ denotes the conditional or transition probability for the system to be at configuration $\mathcal{C}'$ at time t if it was at configuration $\mathcal{C}'$ at time s. The autocorrelation ($C_A(t,s)$) and response ($R_A(t,s)$) functions are defined by,

$$C_A(t,s) = \sum_{\mathcal{C},\mathcal{C}'} P(\mathcal{C}',t|\mathcal{C},s)A(\mathcal{C})A(\mathcal{C}') \qquad ; \qquad R_A(t,s) = \frac{\delta\langle A(t)\rangle}{\delta h(s)} \tag{2}$$

Both C_A and R_A can be decomposed into a stationary (fast) and aging (slow) parts. In the large s regime, for many aging systems a quasi-FDT relates the aging parts of (2) in terms of the fluctuation-dissipation ratio (FDR),

$$R_A^{\rm ag}(t,s) = \frac{1}{T_{\rm eff}^{\rm FDR}(s)} \frac{\partial C_A^{\rm ag}(t,s)}{\partial s}\theta(t-s) \quad . \tag{3}$$

$T_{\rm eff}^{\rm FDR}(s)$ is a time-dependent effective temperature that has been shown to have some of the properties of a thermodynamic temperature [2]. In a weak ergodicity breaking scenario $C_A(t,s)$ decays to zero for $t-s \to \infty$ in a typical time that we denote by $\tau_{\rm decorr}(s)$ [2]. In general, $\tau_{\rm decorr}(s) \propto s^\alpha$ with α a given exponent. In many cases $\alpha = 1$ appears to be a very good approximation, usually termed as simple aging.

We define the aging regime as adiabatic if the fraction of heat released from the system to the bath goes asymptotically to zero for time differences of order of the decorrelation time,

$$\left|\frac{E(s+\tau_{\rm decorr}(s)) - E(s)}{E(s)}\right| \to 0 \qquad \text{for } s \text{ large enough} \tag{4}$$

Of course this relation is meaningful only in the aging regime where the energy $E(s)$ is still far from its equilibrium value [3]. The relation between the rate decay of the energy and the frequency domain where FDT violations are observed has been already pointed out in [4].

3. Stimulated and spontaneous relaxation as the origin of intermittency.

For an adiabatic regime as described in (4) two type of heat exchange processes are predicted to be observable depending on the timescale of the measurement. According to (4), and for times of the order of $\tau_{\rm decorr}(s)$ or smaller, the net heat flow transferred from the system to the bath is exceedingly small, yet heat fluctuations can be as big as if the system were in equilibrium at the bath temperature. In such case there is a continuous heat exchange between the system and the bath and energy fluctuations $< E^2(s) > - < E(s) >^2$ (measured over $\tau_{\rm decorr}(s)$) are of the same order of the energy content $E(s)$ and determined by the heat capacity of the system at the bath temperature.

[2]This could be defined either as the integral correlation time $\int_s^\infty C_A(t',s)dt'$ or the value of $t-s$ for which $C_A(t,s)$ decays to $1/e$ of its maximum value $C_A(s,s)$.

[3]Deviations from the power law divergence $\tau_{\rm decorr}(s) \propto s^\alpha$ may provide an explicit check whether the system has not left the asymptotic regime and aging is not interrupted (in this last case the decorrelation time starts to saturate to its value at equilibrium).

Were one to measure at age s the probability distribution $P_s(Q)$ of heat exchanged Q between the system and the bath along intervals of a given duration $\tau \sim \tau_{\rm decorr}(s)$, a Gaussian distribution would be found with zero mean (as no net heat is transferred from the system to the bath) and a variance $\sigma^2(T)$ which is independent of the age s but dependent on the temperature of the bath. This feature is ubiquitous in glassy matter (e.g. structural glasses quenched at low-enough temperatures) where the net heat flow from the system to the bath is unobservable, yet heat is quickly exchanged with the bath as thermal conductivity is high [4]. For all practical purposes the system looks equilibrated at the temperature of the bath and a thermometer put in contact with it would measure the bath temperature. We call this heat exchange process *stimulated* as it originates in the existence of physical processes thermally excited by the bath.

For times much larger than $\tau_{\rm decorr}(s)$ the average net heat flow from the system to the bath is not negligible, a clear consequence that the system has not yet equilibrated. Most of the exchange processes that occur along these timescales are stimulated by the bath, however other exchange events do not fall into the previous category and follow a completely different distribution. Contrarily to the stimulated process, this distribution is expected to display an exponential tail whose width $\lambda(s)$ depends on the age of the system s. However this process is not excluded to be described by a Gaussian distribution as well. Indeed an exponential tail is characteristic of Gaussian distributions centered around a finite value. We call this new process *spontaneous* as its statistical description is not determined by the temperature of the bath but rather by the fact that the system has been prepared in a non-equilibrium state. Figure 1 illustrates typical distributions for the stimulated and spontaneous processes. The existence of two different kinds of heat exchange process (and therefore two different heat exchange distributions) occurring along widely separated timescales has been recently numerically verified in the context of a simple spin-glass model [5]. It has been suggested that the existence of these two processes is at the roots of the intermittency phenomenon observed in glasses [6] and colloids [7] [5].

4. Statistical (microcanonical) description of the aging state

Along the rest of the paper we will analyze in detail and idealized particular case of the more general previous scenario. In doing this we aim to understand fundamental issues behind a possible statistical, rather thermodynamical, description of the glassy state. We will consider in detail a situation where the stimulated process does not exist and only the spontaneous process is observed in the adiabatic relaxation. We will refer to it as the *partial equilibration* scenario. It is then possible to prove that effective temperatures, with a precise statistical meaning, do emerge. This is the content of the next Subsections, 4.1,4.2,4.3. In Sec. 5 we will focus our attention on the oscillator model (OM) [8] as a simple case where all results of the current Section can be explicitly verified.

[4]The simplest evidence in favor of this statement is a piece of silica vitrified at room temperature (i.e. not yet equilibrated) whose temperature can be felt by touching it with the hand.

[5]For the latter it might seem more appropriate to speak about stress release rather than heat exchanged as compaction (rather than heat dissipation) is the main relaxational process.

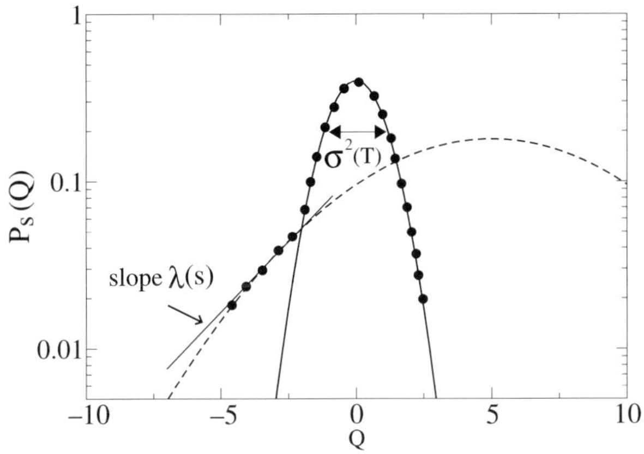

Figure 1. The origin of intermittency: Two mechanisms of relaxation, stimulated (continuous line) and spontaneous (dashed line) characterize the overall relaxation. Q stands for the heat transferred from the bath to the system (the sign indicating the direction of flow, if $Q < 0$ heat flows from the system to the bath) along regularly spaced time intervals and $P_s(Q)$ the corresponding probability distribution at age s. The black points illustrate what experimental measured data would look like. In the simplest scenario both processes are described by a Gaussian distribution. The stimulated component has a variance $\sigma^2(T)$ dependent on the bath temperature but independent on the age. The spontaneous component shows an exponential tail in the $Q < 0$ side (shown as a continuous line) whose slope λ is age dependent. The spontaneous process for $Q > 0$ is unobservable as relaxational dynamics is constrained by the existence of a net heat transfer from the system to the bath.

4.1. Partial equilibration scenario

To suppress the stimulated process the bath is put at zero temperature (therefore, no heat can flow from the bath to the system). However, to keep the system in an adiabatic relaxational regime (4) (avoiding the aging regime to be quickly interrupted due to the presence of forever-lived metastable configurations) we require an additional condition: *Dynamics must be ergodic when the system is constrained to move in any constant energy surface* [6]. This condition is illustrated in Fig. 2. In this case, it is easy to prove that relaxation cannot arrest at $T = 0$ as it is always possible for the system to decrease its energy by moving along the constant energy surface until a favorable downhill move occurs. Models that fall into this category have been termed as models with purely entropic barriers, the OM [8] being a prominent example. Eq. (4) states that in an adiabatic regime the energy remains practically constant during timescales of the order of the decorrelation time $\tau_{\rm decorr}(s)$. Therefore, as correlations decay in timescales much shorter than the energy does, the system can explore a large number of configurations

[6] A more precise statement requires dynamics to be ergodic within the energy shell between E and $E+\Delta E$ where ΔE is finite in the large volume limit.

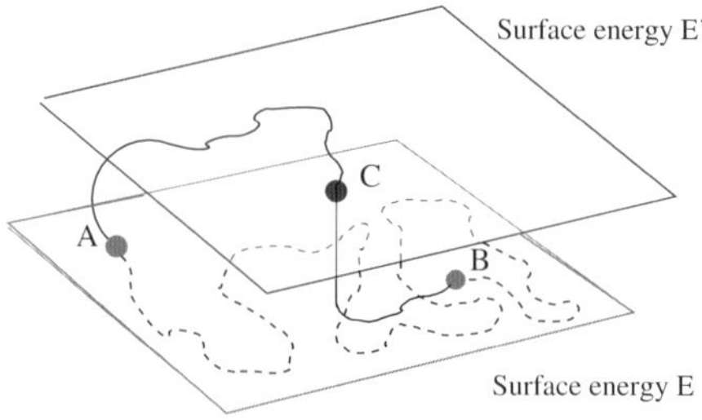

Figure 2. Schematic picture of the type of ergodic dynamics considered in a partial equilibration scenario. Any phase-space points A, B contained in the surface of energy E must be connected by a path lying in that surface (dashed line). Note that other paths (continuous line going through C) that we call non-ergodic (as they change the energy of the trajectory) might also exist as the original dynamics of the system coupled to the bath is ergodic.

within a given region of phase space. In case only spontaneous relaxation takes place, the system is constrained to reach a quasi-stationary state where all configurations lying in a given constant energy surface have the same probability. Equiprobability is guaranteed by the fact that microscopic dynamics over the constant energy surface is ergodic and satisfies detailed balance, i.e. it is reversible over the constant energy surface. The process where a given region of the constant energy surface is sampled according to the microcanonical ensemble (i.e. all configurations lying in that surface have the same probability to be visited) will be termed *partial equilibration*.

For later purposes we define the complexity $\mathcal{S}_E(A)$ by the relation

$$\mathcal{S}_E(A) = \log(\Omega_E(A)) = \log\Big[\sum_{\mathcal{C}} \delta(E - E(\mathcal{C}))\delta(A - A(\mathcal{C}))\Big] \tag{5}$$

where $\Omega_E(A)$ stands for the number of configurations with energy E and observable A [7]. For the total complexity $S(E)$ we have,

$$S(E) = \log(\Omega(E)) = \log\Big[\sum_{\mathcal{C}} \delta(E - E(\mathcal{C}))\Big] \tag{6}$$

where now $\Omega(E)$ is the number of configurations with energy E irrespective of the value of the observable A, $\Omega(E) = \int dA\Omega_E(A)$. In (5) we use the energy label E as a subindex and not as an additional argument, just to stress its key role in the dynamics compared to that of other observables. Note that we could equally well use the subindex s, as for a given age s the average energy of the system is fully determined by the dynamics.

[7] As we are counting configurations, $\mathcal{S}_E(A)$ is nothing more but the usual configurational entropy. However, we deliberately avoid to use this term and prefer to talk about complexity. We have in mind the most general case (not addressed in this paper, see instead [5]) where partial equilibration is established among regions or components of phase space. For this case the use of the term ***configurational entropy*** could be misleading.

4.2. A fluctuation theorem in the aging state

In a partial equilibration scenario transitions between configurations lying at different constant energy surfaces (i.e. transitions that increase or decrease the energy) are not equiprobable. However, after applying an external perturbation, a shift of energy levels takes place and some configurations, initially belonging to different energy surfaces, may end up into the same one. We show below in Sec. 4.3 how the response of the system to an external perturbation is determined by the density of energy levels just around the reference value $E(s)$. Accordingly, transition rates between configurations having different energies E, E', were they equiprobable, are described by the following microcanonical relation,

$$\frac{P(\Delta E)}{P(-\Delta E)} = \frac{\Omega(E')}{\Omega(E)} \tag{7}$$

where $\Delta E = E' - E$ is the (intensive) energy difference between both configurations. It is important to point out that $P(\Delta E)$ is not the actual transition rate, but the rate the system would display if transitions between configurations having different energies were equiprobable, a situation that is encountered only after shifting the energy levels (by applying an external perturbation) and redistributing them into a unique energy surface.

Eq. (7) has the form of a FT since it describes the ratio of rates in the forward $E \to E'$ and the reverse $E' \to E$ directions. Fluctuation theorems, similar to (7), have been derived in other contexts, for instance in systems in steady states [9] or systems arbitrarily perturbed from an initial equilibrium state [10]. From (6) we can write,

$$\frac{P(\Delta E)}{P(-\Delta E)} = \exp\Big[\Big(\frac{\partial S(E)}{\partial E}\Big)\Delta E)\Big] = \exp\Big[\frac{\Delta E}{T_{\rm eff}^{\rm FT}(s)}\Big] \tag{8}$$

where we have expanded the complexity $S(E)$ and kept only the first term in ΔE [8]. The factor $T_{\rm eff}^{\rm FT}(s)$ in the exponent in the r.h.s of (11) defines an effective temperature,

$$\frac{1}{T_{\rm eff}^{\rm FT}(s)} = \Big(\frac{S(E)}{\partial E}\Big)_{E=E(s)} \tag{9}$$

where we specifically include the subindex s to denote its time dependence through the value of the energy E at age s. The super-index FT indicates that this effective temperature is derived from the fluctuation theorem (7). For a Gaussian $P(\Delta E)$ the value of $T_{\rm eff}^{\rm FT}(s)$ can be shown to be proportional to the width of the exponential tail $\lambda(s)$ as depicted in Fig. 1. This connection has been exploited in [5] as a possible way to estimate the effective temperature from intermittency measurements.

Eqs. (7,8) may look as simple detailed balance but it is not. Detailed balance is a inherent property of the microscopic dynamics. Two are the main differences between (8,9) and microscopic detailed balance. In the former we assumed equiprobability between transitions with identical energies and this is not guaranteed if not in the partial equilibration scenario. Moreover the factor (9) in the exponent of (8) is not the temperature of

[8] As ΔE is an intensive quantity, higher order powers in ΔE should be included in (8). However they are not relevant for what is addressed in the present paper. A similar situation is later encountered in (8). See the ensuing footnote (9) for a more detailed explanation.

the bath as implied by detailed balance, but a quantity that is solely determined by the dynamics.

We contend to show that $T_{\text{eff}}^{\text{FT}}(s)$ as derived from the fluctuation theorem coincides with the effective temperature derived from the FDT relations that link correlations and responses in the aging regime. The origin of this connection has been already mentioned. Eq. (7) links transitions between configurations with different energies. These transitions can be probed only by lifting the energy of the different configurations after applying an external field.

4.3. Microcanonical relations and effective temperatures in the aging state.

Let us consider A to be any observable of the system that can take different values in a given surface of constant energy E. The equiprobability assumption implies that the transition rates between configurations with different observable values A, A', at a given age s (when the surface explored has energy E), satisfy the following relation,

$$\frac{W_E(\Delta A)}{W_E(-\Delta A)} = \frac{\Omega_E(A')}{\Omega_E(A)} \tag{10}$$

where $\Delta A = A' - A$ and $\Omega_E(A)$ was defined in (5). The ratio of rates (10) is therefore age dependent as the value of E changes with the age of the system. Eq. (10) says that the rate for the observable A to change its value in ΔA within the surface of constant energy E, when going from the value A to the value A', is proportional to the number of configurations with value A' at the energy E. Although (10) describes rates there is no explicit reference to any timescale. In fact, we use the term W for these rates (as compared with the term $P(\Delta E)$ used in (7)) to stress the fact that these are rates rather than probabilities (i.e. they have dimensions of a frequency). Note that the difference between using P or W is minor as any timescale drops off when considering the ratio among the forward and reverse rates. Again, as was done for (7), we expand the term in the r.h.s. of (10) around $A' = A$ by using (5) and consider only the first term in the expansion,

$$\frac{W_E(\Delta A)}{W_E(-\Delta A)} = \exp\left[\left(\frac{\partial \mathcal{S}_E(A)}{\partial A}\right)\Delta A + \mathcal{O}((\Delta A)^2)\right] \tag{11}$$

The variation $\Delta A = A' - A$ is intensive, therefore all powers of ΔA enter in (11) even in the thermodynamic limit. However, as we will see later, only the first term is relevant for the emergence of effective temperatures. Here we will not discuss the possible relevance of higher order terms [9]. Relation (11) says that dynamics evolves towards configurations that have a higher value of the complexity $\mathcal{S}_E(A)$, i.e. towards the maximum value $A^*(E)$ of $\mathcal{S}_E(A)$,

$$\left(\frac{\partial \mathcal{S}_E(A)}{\partial A}\right)_{A=A^*(E)} = 0 \tag{12}$$

As the value of E is time dependent it is also the value of the maximum $A^*(E)$. In a partial equilibration scenario, A relaxes (along timescales of order of $\tau_{\text{decorr}}(s)$) to the

[9]Higher order terms in ΔA are expected to contain information about the stability of the aging state and thermally induced fluctuations of the effective temperature. Contrarily to the temperature of the bath (which cannot fluctuate) the latter might display fluctuations even in the large V limit.

value $A^*(E)$ given in (12). This value changes in time as the energy decreases and is slaved to the time evolution of the energy. As the energy decreases much slower than A relaxes to $A^*(E)$ (c.f. (4)), reversibility holds in the aging regime,

$$W_E(\Delta A) = W_E(-\Delta A) \tag{13}$$

An interesting and special class of observables are those called *neutral* where, for large enough times, the value of $A^*(E)$ is independent of E (i.e. of the age of the system) $A^*(E) = A^*$. Therefore the value A^* must coincide with the equilibrium value $A^{\rm eq}$ if the system has to equilibrate. Neutral observables have the interesting property that they reach a stationary value exponentially fast and their time evolution is not slaved to that of the energy. The most prominent example of neutral observables is the wave vector density $\rho_k(t)$ in supercooled liquids which stays negligible at all times as there is no long range order that develops in the amorphous glass state. The dynamics of a neutral observable is therefore quite easy to visualize. Starting from any initial configuration the dynamics of the system quickly evolves towards A^* and stays there forever.

Non-neutral observables are expected to relax fast to their value $A^*(E)$ and display an interesting non-monotonic behavior as the value $A^*(E)$ changes with the age s. For instance, if the system starts far from equilibrium but with a value of A corresponding to the equilibrium value $A^{\rm eq}$, initially the system will depart from this value, and follow the time evolution of $A^*(E)$ to eventually come back again to $A^{\rm eq}$ after equilibrating. This effect has been observed in structural glasses where the volume density displays non-monotonic behavior in the glass state and is known as the Kovacs effect [10].

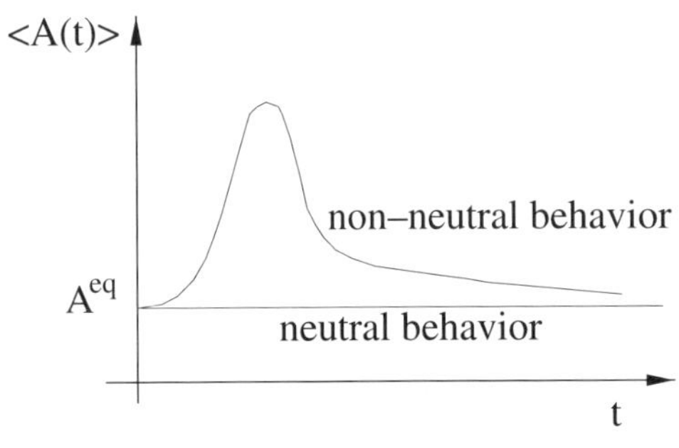

Figure 3. Time evolution of neutral and non-neutral observables. Explanation is given in the text.

We are now in a position to understand the emergence of effective temperatures as usually derived from fluctuation-dissipation relations. Suppose now that at time s an external field of intensity h coupled to the observable A is applied to the system. In the presence of a field the energy of a configuration $\mathcal{C}$ is shifted by the Zeeman term,

$$E(\mathcal{C}) \to E(\mathcal{C}) - hA(\mathcal{C}) \quad . \tag{14}$$

[10]This effect has been studied in different models of glasses, for instance [11,12]

Under the field, the surface of constant energy E does not coincide anymore with that at zero field, meaning that configurations with identical energy at zero field (i.e. lying in a given surface of energy E) get shifted by different amounts in a field (i.e. finish into different surfaces with different values of E). In particular, for a given value of the energy E, from all configurations initially lying in the surface at $h = 0$, after switching the field some configurations leave the surface, others come onto the surface, finally others stay there (i.e. those with $A(\mathcal{C}) = 0$).

According to (14), just after the field has been switched on, configurations with a larger value of A decrease their energy relative to their energy at $h = 0$ and configurations with a lower value of A increase their energy. However, the complexity $S(E)$ (6) is a monotonic function of the energy, therefore configurations with higher energy are more numerous than those with lower energy. As a result of the action of the field, the number of configurations $\Omega_E^h(A)$ with a given value of A, that lie inside the surface of constant energy E, monotonically increases with A,

$$\Omega_E^h(A) = \int dE'\Omega_{E'}(A)\delta(E - E' + hA) = \Omega_{E+hA}(A) =$$
$$\Omega_E(A)\Big[1 + \frac{\partial \log(\Omega_E(A))}{\partial E}Ah + \mathcal{O}(h^2))\Big] = \Omega_E(A)\exp\Big[\Big(\frac{\partial \mathcal{S}_E(A)}{\partial E}\Big)Ah + \mathcal{O}(h^2)\Big] \quad . \tag{15}$$

Using the equivalent of (5), $\mathcal{S}_E^h(A) = \log(\Omega_E^h(A))$, we get,

$$\mathcal{S}_E^h(A) = \mathcal{S}_E(A) + \Big(\frac{\partial \mathcal{S}_E(A)}{\partial E}\Big)Ah + \mathcal{O}(h^2) = \mathcal{S}_E(A) + \beta_{\text{eff}}(E,A)Ah + \mathcal{O}(h^2) \tag{16}$$

where we have defined,

$$\beta_{\text{eff}}(E,A) = \Big(\frac{\partial \mathcal{S}_E(A)}{\partial E}\Big)_{E=E(s),A=A(s)} \tag{17}$$

As $\mathcal{S}_E(A)$ is a monotonic increasing function of E, (15) indicates that relaxation in a field is pushed towards configurations with progressively increasing values of A. In a field partial equilibration occurs again in the new surface of energy E as all configurations there contained remain equiprobable. The new (10) reads,

$$\frac{W_E^h(\Delta A)}{W_E^h(-\Delta A)} = \frac{\Omega_E^h(A')}{\Omega_E^h(A)} = \frac{W_E(\Delta A)}{W_E(-\Delta A)}\exp(\beta_{\text{eff}}(E,A)\Delta A\, h) \tag{18}$$

where (15,16) have been used. A word of caution in the derivation of (18) is in place. We have considered $\beta_{\text{eff}}(E,A) = \beta_{\text{eff}}(E,A')$. This is justified as the quantity $\mathcal{S}_E(A)$ entering into the definition (17) is *extensive*, i.e. proportional to the volume V of the system, while $\Delta A = A' - A$ is an intensive quantity. In the large V limit, the partial derivative (17) is the same whether it is taken at A or A' as the difference $\Delta A = A' - A$ is intensive so $\beta_{\text{eff}}(E,A) - \beta_{\text{eff}}(E,A') \sim \mathcal{O}(1/V)$.

Eq. (18) is reminiscent of detailed balance, however the same remark has to be made here as was done in Sec. 4.2. The quantity $\beta_{\text{eff}}(E,A)$ is not the temperature of the bath anymore but a time dependent value as $E(s)$ and $A(s)$ change in time. The quantity $\beta_{\text{eff}}(E,A)$ defines an effective temperature as obtained from the microcanonical relation (MR) (18),

$$\frac{1}{T_{\text{eff}}^{\text{MR},A}(s)} = \beta_{\text{eff}}(E,A) = \Big(\frac{\partial \mathcal{S}_E(A)}{\partial E}\Big)_{E=E(s),A=A(s)} \quad . \tag{19}$$

As $A(s) = A^*(E(s))$ (see (12) and the ensuing discussion) we can replace the complexity $\mathcal{S}_E(A)$ by the total complexity $S(E)$ in the r.h.s of (19). Notably this gives a value of $T_{\rm eff}^{\rm MR}(s)$ that is independent on the type of observable so the label A in the l.h.s of (19) drops off,

$$\frac{1}{T_{\rm eff}^{\rm MR}(s)} = \beta_{\rm eff}^{\rm MR}(s) = \Big(\frac{\partial \mathcal{S}_E(A))}{\partial E}\Big)_{E=E(s),A=A(s)} = \Big(\frac{\partial S(E)}{\partial E}\Big)_{E=E(s)} \quad , \tag{20}$$

which coincides with the result obtained from the FT (9).

From relation (18) it is now possible to link the autocorrelation and response functions (2) through via the FDR, where the effective temperature (20) appears explicitly,

$$R_A(t,s) = \frac{1}{T_{\rm eff}^{\rm MR}(s)}\frac{\partial C_A(t,s)}{\partial s}\theta(t-s) \quad . \tag{21}$$

We are not going to show here the details of this derivation but only mention the main steps. The proof follows the scheme of the derivation shown in Section 3 of Ref. [3] for the standard derivation of FDT where (18) here corresponds to (49) there. In [3] it is shown how the response function can be decomposed in two different contributions called $R^{(1)}$ and $R^{(2)}$. In equilibrium, when FDT holds, $R^{(1)}$ vanishes and $R^{(2)}$ gives the equilibrium response. In the partial equilibration scenario, (21) is obtained whenever $R^{(1)}$ vanishes. Close inspection of (53) in [3] shows that this term vanishes only if the perturbing field does not modify, to linear order in h, the trapping time distribution measured over transitions starting at the surface of energy E. More precisely, consider a sample of all trapping times for configurations of age equal to s. As dynamics is stochastic, the same quenching protocol will generate different configurations with values E, A in the vicinity of their dynamical averages $E(s), A(s)$. Let $p_s(\tau), p_s^h(\tau)$ be the trapping time distributions both at zero field and after applying a field h at time s respectively. Eq. (21) holds whenever $p_s^h(\tau) = p_s(\tau) + \mathcal{O}(h^2)$ [11]. In other words, up to linear order in h the sole effect of the field is to modify the density of configurations as indicated in (15).

The physical significance of these results can now be appreciated at its full extent. In the simplest partial equilibration scenario the effective temperature derived from the microcanonical relation (20) is independent on the type of observable and coincides with that obtained from the FT (9) and the FDR (3). For the last equality, the trapping time distribution is required to remain unchanged at linear order with the intensity of the field h. In this case,

$$T_{\rm eff}^{\rm FDR}(s) = T_{\rm eff}^{\rm FT}(s) = T_{\rm eff}^{\rm MR}(s) \tag{22}$$

This equality is remarkable as it shows that the effective temperature can be directly obtained from the fluctuation theorem (9) without the necessity of introducing a perturbing field and measuring correlations and responses [12]. The first equality in (22) has been numerically verified in a given example of spin-glass model [5]. In what follows we exemplify these results for a simple solvable case.

[11] Similarly one could say that the average trapping time is not modified to linear order in h

[12] From a different perspective, a result where a perturbing field is not required to measure the FDR has been recently proposed [13,14].

5. The oscillator model (OM)

Exactly solvable oscillator models (OMs) [8], unrealistic as they look, provide a simple physical basis to describe glassy dynamics. A parallel can be established between oscillator models for glassy dynamics and the original urn models by the Ehrenfest's. The former can enlighten our comprehension of the essential features of glassy dynamics, in the same way urn models have provided a ground basis to understand key concepts of equilibrium thermodynamics such as the Boltzmann entropy. The OM is an example where the partial equilibration scenario holds for the dynamics at $T = 0$. It provides an excellent framework to verify the statements of previous sections.

5.1. Energy relaxation

The OM [8] is described by a set of continuous variables x_i and an energy function,

$$E = \frac{K}{2}\sum_{i=1}^{N} x_i^2 \tag{23}$$

where K is the spring constant and N is the total number of oscillators. Oscillators are non-interacting and therefore the model has trivial static properties, the total energy $E = Nk_BT/2$ according to the equipartition law. We consider a cooperative dynamics where oscillators are updated according to the rule $x_i \to x_i + \frac{r_i}{\sqrt{N}}$, the r_i being uncorrelated Gaussian variables with $\overline{r_i} = 0$ and variance $\overline{r_i r_j} = \Delta^2\delta_{ij}$. The updating of all oscillators is carried out in parallel and the moves are accepted according to the Metropolis rule. We will focus our analysis on the dynamics at $T = 0$ where only updates that decrease the total energy are accepted. This leads to slow dynamics as most of the proposed moves tend to increase the energy and only few of them are accepted (i.e. the acceptance rate is quickly decreasing with time). Dynamics in the OM is ergodic if confined to a constant energy surface (see Fig. 2). Therefore, dynamics does not become arrested at $T = 0$ as no metastable configurations (except the ground state) exist at $T = 0$.

Dynamical properties in the OM are derived from the distribution of attempted energy changes $P(\Delta E)$. There are several ways to compute this distribution [15], the simplest one derives from the Gaussian character of such distribution. The change in energy of an elementary move is given by

$$\Delta E = \frac{K}{\sqrt{N}}\sum_i x_i r_i + \frac{K}{2N}\sum_i r_i^2. \tag{24}$$

From the Gaussian character the of r_i, it follows that ΔE has a Gaussian distribution whose mean and variance are given by,

$$M_{\Delta E} = \overline{\Delta E} = K\Delta^2/2, \qquad \sigma_{\Delta E} = \overline{(\Delta E)^2} - (\overline{\Delta E})^2 = 2K\Delta^2\frac{E}{N} = 2K\Delta^2 e \tag{25}$$

which yields [8],

$$P(\Delta E) = (4\pi e K\Delta^2)^{-\frac{1}{2}} \exp\left[-\frac{(\Delta E - \frac{K\Delta^2}{2})^2}{4eK\Delta^2}\right] \tag{26}$$

where $e = E/N$ is the energy per oscillator. At $T = 0$, the dynamical evolution of the energy e and the acceptance rate $a(e)$ (the fraction of accepted moves) are given by the following closed equations,

$$\frac{\partial e}{\partial t} = \int_{-\infty}^{0} xP(x)\,dx \quad ; \quad a(e) = \int_{-\infty}^{0} dx P(x) = \frac{1}{2}\mathrm{Erfc}\big(\sqrt{\frac{K\Delta^2}{16e}}\big) \tag{27}$$

with $\mathrm{Erfc}(x) = (2/\sqrt{\pi})\int_x^\infty du \exp(-u^2)$ the complementary error function. As the energy e decreases the variance of the distribution (25) decreases. Accordingly, the acceptance rate also decreases. The dynamical evolution of the energy and acceptance can be solved in the long-time limit, $E(t) \sim 1/\log(t), a(t) \sim 1/(t\log(t))$.

5.2. Effective temperature

Correlations and responses have been computed for the magnetization $M = \sum_i x_i$ [8]. In the rest of the paper we will consider $A = M$ as the observable under quest. M is a neutral observable as can be verified by solving the dynamical equation for the magnetization. It relaxes exponentially fast to zero which is the equilibrium value of the magnetization at all temperatures. The autocorrelation function $C_M(t,s) = (1/N)\sum_i x_i(t)x_i(s)$, the corresponding response function $R_M(t,s)$ (2) and the susceptibility $\chi_M(t,s) = \int_s^t R_M(t,t')dt'$, do not have stationary part but only aging part and show simple aging with $\tau_{\mathrm{decorr}}(s) \propto s$. Correlations and responses define an effective temperature $T_{\mathrm{eff}}(t,s)$ through the relation,

$$T_{\mathrm{eff}}^{\mathrm{FDR}}(t,s) = \frac{\frac{\partial C_M(t,s)}{\partial s}}{R_M(t,s)} \tag{28}$$

which is the temperature T of the bath in equilibrium. At $T = 0$ the response is finite in the OM due to the shift of energy levels described in Sec. 4.3. A simple expression can be derived in that case,

$$T_{\mathrm{eff}}^{\mathrm{FDR}}(t,s) \equiv T_{\mathrm{eff}}^{\mathrm{FDR}}(s) = 2e(s) + \frac{1}{f(s)}\frac{\partial e(s)}{\partial s} \tag{29}$$

where $f(s)$ is a given function [8] which asymptotically decays as $1/s$. Two remarkable facts emerge from (29): 1) $T_{\mathrm{eff}}(t,s)$ only depends on the lowest time s, therefore characterizes the aging state of the system at time s; 2) The second term in the r.h.s of (29) is sub-dominant respect to the first term leading to $T_{\mathrm{eff}}(s) \to 2e(s)$, i.e. the equipartition law is asymptotically satisfied in the aging regime.

5.3. The fluctuation theorem

As the energy decays logarithmically and $\tau_{\mathrm{decorr}}(s) \propto s$, (4) is satisfied and relaxation is adiabatic. Moreover, dynamics in this model is ergodic if constrained to the constant energy surface. Under these conditions, only spontaneous relaxation occurs (Sec. 3) and partial equilibration holds (Sec. 4.1). We can verify the validity of the FT by substituting (26) in (8),

$$\frac{P(\Delta E)}{P(-\Delta E)} = \exp\big(\frac{\Delta E}{T_{\mathrm{eff}}^{\mathrm{FT}}}\big) = \exp\big(\frac{\Delta E}{2e}\big) \tag{30}$$

giving $T_{\mathrm{eff}}^{\mathrm{FT}} = 2e$. This result coincides with the asymptotic value $T_{\mathrm{eff}}^{\mathrm{FDR}}$ previously obtained (29).

5.4. Trapping time distribution

Here we compute the trapping time distributions without and in a field $p_s(\tau), p_s^h(\tau)$. In the presence of a field the energy of the OM is given by,

$$E = Ne_T = \frac{K}{2}\sum_{i=1}^{N} x_i^2 - h\sum_{i=1}^{N} x_i = N(e - hm) \qquad (31)$$

where e, m denote $(K/2)\overline{x^2}$ and $\overline{x}$ (i.e. the energy and magnetization per oscillator respectively), the total energy per oscillator being $e_T = e - hm$. Eq. (31) can be written as $E = \frac{K}{2}\sum_{i=1}^{N}(x_i - h/K)^2 - Nh^2/(2K)$. The updating rule for the shifted variables $y_i = x_i - h/K$ remains unchanged. Consequently, the evolution equation for the quantity $e' = (K/2)\overline{y^2} = e - hm + h^2/(2K) = e_T + h^2/(2K)$ is identical to that obtained for e at $h = 0$. As the time evolution of e' does not depend on h this implies that e_T gets corrections that are even powers of h. In a field, (27) holds by replacing e with e'. Therefore, the acceptance rate is not modified at linear order in h. In general, the trapping time probability distribution is given by $p_s^h(\tau) = a^h(e')(1 - a^h(e'))^{\tau-1}$ where τ is the finite number of Monte Carlo steps (i.e. does not scale with N) and $a^h(e')$ is the acceptance rate in a field. The same expression is valid for $p_s(\tau)$ putting $h = 0$. This immediately proves that the distribution of trapping times remains unchanged at linear order in h. In particular, the average trapping time is finite and given by the inverse of the acceptance rate,

$$\overline{\tau^h(e')} = \frac{\tau_0}{\int_{-\infty}^{0} P_h(\Delta E)d(\Delta E)} = \frac{\tau_0}{a^h(e')} \qquad (32)$$

where τ_0 is a microscopic time.

5.5. Microcanonical rates for the magnetization

Let us consider the joint probability $P^h(\Delta E, \Delta M)$ of having a change in the total energy ΔE and magnetization ΔM in the presence of an external field h (E is given in (31) and includes the Zeeman term). At $T = 0$, the probability a_E^h that an attempted change ΔM is accepted is given by,

$$W_E^h(\Delta M) = \frac{1}{\tau_0}\int_{-\infty}^{0} P^h(\Delta E, \Delta M)d(\Delta E) = \frac{1}{\tau_0}P^0(\Delta M)\int_{-\infty}^{0} P_c^h(\Delta E|\Delta M)d(\Delta E) \qquad (33)$$

where $P^h(\Delta E, \Delta M) = P_c^h(\Delta E|\Delta M)P^0(\Delta M)$, the subscript c standing for conditional probability [13]. Again, it is easy to show that these distributions are all Gaussian. Straightforward calculations give,

$$P^0(\Delta M) = (2\pi\Delta^2)^{-\frac{1}{2}}\exp\Big[-\frac{(\Delta M)^2}{2\Delta^2}\Big] \qquad (34)$$

$$P_c^h(\Delta E|\Delta M) = (2\pi\alpha_h^2)^{-\frac{1}{2}}\exp\Big[-\frac{(\Delta E - \frac{K\Delta^2}{2} - b_h\Delta M)^2}{2c_h^2\sigma_h^2}\Big] \qquad (35)$$

with the definitions $\alpha_h = c_h\sigma_h$, $b_h = g_h\sigma_h^2$, $\sigma_h^2 = 2Ke + h^2 - 2Khm$, $c_h^2 = \Delta^2(1 - \frac{(Km-h)^2}{\sigma_h^2})$, $g_h = (Km - h)/(2Ke + h^2 - 2Khm)$. In the linear response regime we obtain, $g_h =$

[13] We are following the notation of Sec. 4.3 where $W_E(\Delta M)$ stands for the rate at zero field.

$\frac{m}{2e}+(m^2/(2e^2)-1/(2Ke))h+\mathcal{O}(h^2)$, $c_h^2=c_0^2(1+mh/e+\mathcal{O}(h^2))$, $c_0^2=\Delta^2(1-m^2K/(2e))$, $\sigma_h^2=2Ke(1-hm/e+\mathcal{O}(h^2))$,$b_h=mK-h+\mathcal{O}(h^2)$. Inserting (34,35) in (33) gives,

$$W_E^h(\Delta M)=W_E(\Delta M)\exp\Big[\frac{(\Delta M)h}{4e-2m^2K}(1+\frac{2m\Delta M}{\Delta^2})\Big] \tag{36}$$

$$W_E(\Delta M)=\frac{1}{\tau_0}\frac{P^0(\Delta M)}{2}\mathrm{Erfc}\Big[\frac{(k\Delta^2)/2+b_0(\Delta M)}{2^{1/2}\alpha_0}\Big] \tag{37}$$

where $P^0(\Delta M)$ is given by (34). As m is a neutral observable that relaxes fast to 0 we can replace $m=0$ everywhere in all previous expressions. Up to linear order in h the rates W_E^h are given by,

$$W_E^h(\Delta M)=W_E(\Delta M)\exp\Big(\frac{\Delta Mh}{4e}\Big) \quad ; \quad W_E(\Delta M)=P^0(\Delta M)a(e) \tag{38}$$

where $a(e)$ is given in (27). We remark the following results: 1) the perturbed rates are multiplicative [14], 2) (13) holds as $W_E(\Delta M)=W_E(-\Delta M)$ and 3) relation (18) is satisfied as well,

$$\frac{W_E^h(\Delta M)}{W_E^h(-\Delta M)}=\frac{W_E(\Delta M)}{W_E(-\Delta M)}\exp\Big[\frac{(\Delta M)h}{2T_{\rm eff}^{\rm MR}(e)}\Big] \tag{39}$$

so the effective temperature here obtained $T_{\rm eff}^{\rm MR}(e)=2e$ again coincides with that derived from the FDR (29) and the FT (30).

6. Conclusions

A statistical interpretation of a non-equilibrium or effective temperature in aging systems, rather than a thermometric one, might be possible. Recent numerical simulations of a disordered model [5] have shown that the effective temperature, as measured from FDT violations, originate from the existence of a *spontaneous* relaxational process describing heat exchange low-frequency events. Superimposed to it there is a *stimulated* process that is characterized by the temperature of the bath and responsible of most of the heat exchange observable events between system and bath. Although the timescale of the spontaneous process is related to the temperature of the bath, its statistical description (e.g. the specific form of the corresponding FT) is related to other properties of the aging state such as the energy content. The concurrence of these two processes in the overall relaxation is related to the intermittent phenomenon recently observed in various experiments [6,7]. Moreover, the width of the exponential tail associated to the spontaneous process is predicted to be proportional to the effective temperature.

We have investigated a partial equilibration scenario where only the spontaneous process occurs (the temperature of the bath is set to zero) and dynamics is ergodic when constrained to a given energy surface. In this case three effective temperatures can be

[14]The form of the perturbed rates (38) is multiplicative. This assumption was made in a given class of trap models at the level of configurations and shown to lead to the existence of effective temperatures [16]. Although the multiplicative property may hold at the coarse-grained level of observable values, it was shown to be far-fetched at the level of configurations [17]. Eq. (38) shows that rates are to be considered multiplicative only at the level of observable values.

defined: a) from the fluctuation-dissipation ratio (3); b) from the fluctuation theorem for the statistical description of the spontaneous component (7); c) from microcanonical relations relating observable changes (20). All three are shown to be identical (22). Explicit calculations in the OM demonstrate the validity of these statements.

Several open questions and directions of research along the present ideas appear worthwhile. It would be useful to go beyond the qualitative level of demonstrations in the present paper over more founded mathematical proofs of all these results. Attempts trying to establish the origin of effective temperatures using master equation formalisms have already appeared in the literature [13,18] and we foresee more work in the near future. We should also mention the close similarity between the present approach and that by Edwards for granular matter [19]. Would be very interesting to investigate other classes of models with simple equilibrium properties, such as kinetically constrained models [20], where the existence of non-trivial effective temperatures is still under debate. In these models it is possible to identify correlated motion of particles [21] that underpin a spatially heterogeneous dynamics [22]. Another category of interesting problems to explore are systems in steady states where FDT violations have been studied [23] and where non-Gaussian effects, similar to those described here, have been identified as well [24].

REFERENCES

1. J. Casas-Vazquez and D. Jou, Rep. Prog. Phys. **66**, 1937 (2003).
2. L. F. Cugliandolo, J. Kurchan and L. Peliti, Phys. Rev. E **55**, 3898 (1997).
3. A. Crisanti and F. Ritort, J. Phys. A (*Math. Gen.*), **36**, R181-R290 (2003).
4. L. F. Cugliandolo, D. S. Dean and J. Kurchan, Phys. Rev. Lett **79**, 2168 (1997).
5. A. Crisanti and F. Ritort, Preprint **cond-mat/0307554**
6. L. Buisson, L. Bellon and S. Ciliberto, J. Phys. C (*Cond. Matt.*) **15**, S1163 (2003)
7. L. Cipelletti et al., J. Phys. C (*Cond. Matt.*) **15**, S257 (2003)
8. L. L. Bonilla, F. G. Padilla and F. Ritort, Physica A, **250**, 315 (1998).
9. D. Evans and Searles, Adv. Phys. **51**, 1529 (2002).
10. G. E. Crooks, J. Stat. Phys. **90**, 1481 (1998).
11. E. M. Bertin, J. P. Bouchaud, J. M. Drouffe and C. Godreche, Preprint **cond-mat/0306089**
12. A. Buhot, Preprint **cond-mat/0310311**
13. C. Chatelain, J. Phys. A (*Math. Gen.*), **36**, 10739 (2003)
14. F. Ricci-Tersenghi, Preprint **cond-mat/0307565**
15. A detailed exposition can be found in Sec. 6.5.1 of [3].
16. F. Ritort, J. Phys. A (*Math. Gen.*), **36**, 10791 (2003)
17. P. Sollich, J. Phys. A (*Math. Gen.*), **36**, 10807 (2003)
18. G. Diezemann, Preprint **cond-mat/0309105**
19. S. F. Edwards, Physica A, **157**, 1080 (1898).
20. F. Ritort and P. Sollich, Adv. Phys. **52**, 219 (2003)
21. J. P. Garrahan and D. Chandler, Proc. Nat. Acad. Sci. **100**, 9710 (2003)
22. L. Berthier and J. P. Garrahan, Preprint **condmat/0306469**
23. I. Santamaría-Holek, D. Reguera and J. M. Rubí, Phys. Rev. E **63**, 051106 (2001)
24. R. van Zon and E. G. D. Cohen, Preprint **cond-mat/0310265**

Unifying Concepts in Granular Media and Glasses
A. Coniglio, A. Fierro, H.J. Herrmann and M. Nicodemi (editors)

Heterogeneities in the Dynamics of Supercooled Water

Nicolas Giovambattista[a], Marco G. Mazza[a], Sergey V. Buldyrev[a], Francis W. Starr[b] and H. Eugene Stanley[a]

[a]Center for Polymer Studies and Department of Physics
Boston University, Boston, MA 02215 USA

[b]Polymers Division and Center for Theoretical and Computational Materials Science
National Institute of Standards and Technology, Gaithersburg, MD 20899 USA

We investigate the dynamics of liquid water by performing molecular dynamics (MD) simulations using the SPC/E model. We identify clusters of mobile molecules indicating the presence of spatially heterogeneous dynamics. We study the time and temperature dependence of the cluster size and find that clusters are larger at the time when molecules start to diffuse and grow as temperature decreases. We relate the average mass n^* of mobile particle clusters to the diffusion constant and the configurational entropy. We find that n^* can be interpreted as the mass of the "cooperatively rearranging regions" that form the basis of the Adam-Gibbs theory of the dynamics of supercooled liquids. In the context of the potential energy landscape (PEL) approach, the diffusion of molecules is related to change of basins. By studying the dynamics of the system on the PEL we identify clusters formed by the molecules with larger displacements between neighbor basins. These transitions between basins of the PEL are the elementary steps of diffusion and therefore of the spatially heterogeneous dynamics observed in the real MD-trajectory.

1. Introduction

Supercooled liquids are characterized by the non-exponential decay of (ensemble-averaged) time correlation functions [1–3] and usually such a decay can be expressed in terms of a stretched exponential function, $\exp\left[-(t/\tau)^{\beta}\right]$ [4]. Two microscopic scenarios have been proposed to explain this behavior; they are schematically shown in Fig. 1.

In the "homogeneous" scenario, correlation functions for different molecules decay in the same way, e.g., by a unique stretched exponential function with a characteristic relaxation time τ and exponent β. Instead, in the "heterogeneous" scenario, correlation functions for different molecules decay exponentially with a distribution of relaxation times. The superposition of these individual exponential contributions produces a non-exponential decay of the (ensemble-averaged) time correlation function, the exponent β being a measure of the width of the distribution of relaxation times.

As shown in Fig. 1, in the "homogeneous" scenario all molecules are equivalent and as temperature is lowered the locally averaged molecule displacement is the same at every point in the system box. Instead, in the "heterogeneous" scenario, the locally averaged molecule displacements are different depending on the part of the system box we are looking at. One finds groups of molecules that are more "mobile" and groups that are less mobile than the average molecule in the system. Furthermore, the more mobile molecules form patches (indicating a cooperative motion) that increase in size as the temperature is lowered. These patches of mobile molecules have a short lifetime and appear and disappear constantly in different parts of the system.

Recent experiments [5–10] and simulations [11–14] have identified dynamical heterogeneities in supercooled liquids and spin glasses [15]. In Sec. 2 we show that spatially heterogeneous dynamics aptly describe the dynamics in supercooled water.

A successful theoretical approach to the study of supercooled liquids is the potential energy landscape (PEL) approach. The PEL for a system composed of N particles is the potential energy surface of a system in a (3N+1)-dimension configurational space. The system in this high-dimensional space is represented by a point to which coordinates are given by the coordinates of the particles. The diffusion of the particles in the system correspond to the motion of the representative point of the system on the PEL. [16–21]. As a liquid is cooled toward the glassy state, the ability of the particles to diffuse decreases and the system is increasingly found near local potential energy minima on the PEL, called inherent structure (IS) configurations [17]. In this description, in the glassy state, the system is localized in one of the potential energy basins [22–25]. A general picture of the system moving among a set of basins surrounding the multitude of local minima has evolved. A connection between PEL properties and dynamics has been already proposed by Adam-Gibbs (AG) [26] and confirmed in computer simulations using the SPC/E model [27]. However, the AG theory is based on the concept of cooperative rearranging regions (CRR), which is not precisely defined. In section 3, we relate the SHD-clusters we find in our simulation with the CRR from the AG theory.

The PEL approach proposes that diffusion occurs by IS-transitions, i.e., by moving from one basin to another. Furthermore, as Stillinger suggests [18], these IS-transition involve more subtle heterogeneities corresponding to rearrangement of localized set of molecules. In Sec. 4 we identify these more subtle heterogeneities associated to the IS-transitions. Finally, in Sec. 5 we discuss our results.

2. Spatially Heterogeneous Dynamics

Clusters composed of particles with high mobility have been found in numerical simulations of simple systems (e.g., Lennard-Jones mixtures); indicating the presence of spatially heterogeneous dynamics (SHD) [28–36]. Hence, the heterogeneous scenario to describe the dynamics of cold liquids was confirmed. In this section we show that SHD are also present in computer simulations of the SPC/E [37] water model. We study a system with $N = 1728$ molecules at fixed density $\rho = 1.0$ g/cm^3 varying the temperature T from 200 K to 260 K in steps of 10 K. In order to increase statistics we performed two independent simulations for every temperature (in agreement with Ref. [4]). We find that the $T-$dependence of the diffusion constant can be expressed by:

$$D \sim (T - T_{MCT})^{\gamma} \tag{1}$$

where the mode coupling temperature $T_{\text{MCT}} = 193$ K and the diffusivity exponent $\gamma = 2.80$.

As already used in a study of a Lennard-Jones (LJ) mixture [30] and experiments on colloids [36], we use the same approach to define SHD clusters. Let t^* be the value of time for which the non-Gaussian parameter

$$\alpha_2(t) \equiv \frac{3}{5}\langle r^4(t)\rangle / \langle r^2(t)\rangle^2 - 1 \tag{2}$$

presents a maximum [1]. We will see that t^* is the average time for a molecule to diffuse. We then calculate the self part of the time-dependent van Hove correlation function [38] $G_s(r,t)$ *at* $t = t^*$,

$$G_s(r,t^*) = \frac{1}{N}\sum_{i=1}^{N} \langle \delta(|\vec{r}_i(t^*) - \vec{r}_i(0)| - r)\rangle \tag{3}$$

where $\langle\rangle$ represents average over configurations and $\vec{r}_i(t)$ the coordinates of the oxygen atom of the *i-th* molecule; the probability at time t of finding an oxygen atom at a distance r from its position at $t = 0$ is given by $4\pi r^2 G_s(r,t)dr$. We also fit $G_s(r,t)$ with a Gaussian approximation $G_0(r, t = t^*)$ (see Fig. 2),

$$G_0(r,t^*) = \left[\frac{3}{2\pi\left\langle r^2(t^*)\right\rangle}\right]^{3/2} \exp\left[-3r^2/2\left\langle r^2(t^*)\right\rangle\right] \tag{4}$$

where $\langle r^2(t^*)\rangle$ is the mean square displacements of the oxygen atoms *at* $t = t^*$. For very short times $G_s(r, t \to 0) \approx G_0(r, t \to 0)$. However as time increases, deviations of $G_s(r,t)$ from a Gaussian distributions are more pronounced.

In Fig. 2, we see that these distributions intersect for large r at r^*, and that $G_s(r,t^*)$ develops a tail for large r falling outside the Gaussian distribution. Molecules which displacements $r > r^*$ can be considered as molecules that move more than expected (in comparison to $G_0(r,t^*)$). We find r^* is in the range $0.2 - 0.25$ nm for all T (the oxygen-hydrogen distance in a molecule for SPC/E is 0.1 nm). If there are SHD, one would expect molecules with displacements $r > r^*$ to get involved in the SHD clusters. The fraction of molecules with $r > r^*$ at $t = t^*$ is given by $\phi \equiv \int_{r^*}^{\infty} 4\pi r^2 G_s(r,t^*)dr$. Depending on T, we find $6\% < \phi < 8\%$. For simplicity we fix $\phi = 7\%$ for all T. Similar values of ϕ were found in atomic systems [29,30,36] and in polymer melts [39]. Following [30,36], we define the mobility of the *i-th* molecule at a given time t_0 as the maximum displacement of the oxygen atom in the interval $[t_0, t_0 + \Delta t]$,

$$\mu_i(t_0, \Delta t) = max\{|\vec{r}_i(t_0) - \vec{r}_i(t + t_0)|, t_0 \le t \le t_0 + \Delta t\} \tag{5}$$

and we will be interested on the "mobile" molecules defined as the fraction ϕ of molecules with larger μ_i. Finally, we define a SHD-cluster for an observation time Δt as those mobile molecules whose nearest neighbor oxygen-oxygen distance at time t_0 is less than 0.315 nm, the first minimum of the oxygen-oxygen radial distribution function [40].

We find in water similar SHD present in models of simpler liquids. Fig. 3 shows two snapshots of mobile particle clusters at $T = 260$ K for $\Delta t = t^*$. In LJ systems [29], mono-atomic liquids [41], and polymers [42] complex clusters are composed of more elementary "strings" in which particles are arranged in a roughly linear fashion. This is not so clear in the case of our clusters because the hydrogen bond network constrains the geometry of the clusters.

2.1. Δt- and t-dependence of Cluster Size

We first address the issue of the dependence of SHD clusters on the observation time Δt. The quantities we study are $\langle n(\Delta t)\rangle$ and $\langle n(\Delta t)\rangle_w = \langle n^2(\Delta t)\rangle / \langle n(\Delta t)\rangle$; $\langle n(\Delta t)\rangle_w$, the weight average cluster size, is the average size of a cluster to which a randomly chosen molecule belongs. Figure 4 shows $\langle n(\Delta t)\rangle$ and $\langle n(\Delta t)\rangle_w$ for $T = 210$ K. To eliminate the random contribution, we normalize $\langle n(\Delta t)\rangle_w$ by $\langle n_r\rangle_w$, the weight average cluster size for ϕN randomly chosen molecules. We also show in Fig. 4(a), for comparison, the non-Gaussian parameter $\alpha_2(\Delta t)$ and the mean-squared displacement $\langle r^2(\Delta t)\rangle$. The three characteristic time regimes: ballistic, cage and diffusive are indicated in Fig. 4(a)

We find that $\langle n(\Delta t)\rangle_w / \langle n_r\rangle_w$ behaves in a way similar to polymer systems [39] but differs in that there is a clear increase in $\langle n(\Delta t)\rangle_w / \langle n_r\rangle_w$ at the time scale at which molecules go from a ballistic to a cage regime. We attribute this to strong correlations in the vibrational motion of the first-neighbor molecules, owing to the presence of hydrogen bonds.

In Fig. 4(c) we show $\langle n(\Delta t)\rangle_w / \langle n_r\rangle_w$ for all considered T. For $T \leq 240$ K, the maximum in $\langle n(\Delta t)\rangle_w / \langle n_r\rangle_w$ increases in magnitude and shifts to larger time scales with decreasing T. The plateau at the crossover from the ballistic regime is nearly T-independent, as expected since the mean collision time is nearly T-independent. For $T \geq 250$ K, the maximum and the plateau merge, and hence it is not possible to separately distinguish these features.

As we mentioned in the introduction, in the heterogeneous scenario clusters of mobile molecules appear and disappear constantly with time. In Fig. 5, we show the time evolution of a cluster defined using $\Delta t = t^* \approx 65$ ps, at $T = 210$ K. We define $t = 0$ as the time at which the cluster is defined. The corresponding snapshot is shown in the upper left panel. The subsequent snapshots for increasing values of time are shown in Fig. 5 from left-to-right and top-to-bottom. The times of each snapshot are indicated schematically by arrows at the bottom of the figure. Times separating the three regimes (ballistic, cage and diffusive) are also identified. During the ballistic regime and until the beginning of the cage regime, we observe no change in the cluster structure. During the cage regime the collisions of the molecules with the neighbor molecules produce little effect. At $t \approx 16.4$ ps only one molecule of the cluster (at the right end) change hydrogen bonds. These cumulative little effects produce a split of the cluster as time reaches the end of the cage regime. For our particular case, at the beginning of the diffusive regime molecules are able to re-combine forming two residual sub-clusters. As expected, molecules for longer times diffuse farther and farther from each other. No memory of the starting structure of the cluster remains. During this process other clusters appear and disappear in the system.

2.2. T-dependence of Clusters

We focus now on the temperature dependence of the clusters obtained at $\Delta t = t^*$, (when clusters are larger). Because we used the same definition of clusters as in Ref.[30] we can compare our results with those found there for a LJ mixture. Fig. 6 shows the probability distribution $P(n, T)$ to find a cluster with n molecules for different temperatures T using a fraction $\phi = 0.07$. We note that there are not percolating clusters for this system size and $\phi = 0.07$. The distributions behave as $P(n, T) \sim n^{-\tau(T)} \exp[-n/n_0(T)]$. The parameters in this expression are tabulated in Table I. Changing the value of ϕ does not affect the functional dependence of $P(n,T)$ although it changes the values of $n_0(T)$ and $\tau(T)$. The values of $n_0(T)$ do not follow a simple dependence on T. However, $n_0(T)$ which is a characteristic size at T indicate that clusters get larger when cooling. From Table I clearly $\tau(T) \approx 2$ for all T. This value coincides with the value obtained for LJ particles ($\tau(T) \approx 1.9$) [30] and for colloids ($\tau(T) \approx 2.2 \pm 0.2$) [36]. A similar expression for $P(n,T)$ was also found in [39] for a polymer melt. Clusters found in water are much smaller than those found in LJ systems. In Ref.[30] it was found that $P(n,T)$ for $T = 1.07\ T_{MCT}$ and $\phi = 0.05$ is non-zero up to $n \approx 80$ while from Fig. 6 we see that even at $T = 200$ K (i.e., 1.03 T_{MCT}) there are not clusters approximately for approximately $n > 50$. The difference is more clear if we compare $P(n,T)$ in both systems at higher temperatures.

3. SHD and the Adam-Gibbs Theory

Many years ago Adam and Gibbs (AG) proposed a theory to describe the dynamics of supercooled liquids[26,43]. They introduced the concept of "cooperatively rearranging regions" (CRR) to describe the diffusion in a cold liquid. Clearly the AG theory assumes that the heterogeneous scenario is correct.

The theory predicts the empirical Williams-Landel-Ferry equation which determines the evolution of the relaxation time with temperature. Another important result is the relation between the diffusion constant D, the temperature T and the configurational entropy of the system $S_{\rm conf}$,

$$D \propto \exp(-A/TS_{\rm conf}). \tag{6}$$

$S_{\rm conf}$ is interpreted, in the thermodynamic limit, as $k_B \log W_c$, where W_c is the number of configurations accessible to the system and k_B is the Boltzmann constant. In a more modern interpretation W_c has been identified as the number of basins in the PEL accessible to the system in equilibrium, and this allows an easier direct calculation of $S_{\rm conf}$ by computer simulations[44]. Eq. 6 has been tested and appears to be valid across a wide spectrum of liquids [27,45].

The AG theory has been successfully tested, however it is based on the vague definition of CRR. In their work, AG define a CRR as "...a subsystem of the sample which, upon a sufficient fluctuation in energy (or, more correctly, enthalpy), can rearrange into another configuration independently of its environment". There is no quantitative definition of CRR, however, AG predict that the characteristic mass z of the CRR is related to the configurational entropy of the CRR $s_{\rm conf}(z)$ and the total configurational entropy $S_{\rm conf}$ by,

$$z = N s_{\rm conf}(z)/S_{\rm conf} \tag{7}$$

where N is the number of molecules in the liquid. Based on this expression, we will propose a quantitative definition of CRR in the context of the SHD analysis described in the section above.

In order to find a relation between SHD and AG predictions, we calculate the average cluster mass $\langle n(\Delta t)\rangle$ for each T for $\Delta t = t^*$. Motivated by the recent results that the average instantaneous cluster mass scales inversely with the entropy in a model of living polymerizations [46] and based on Eq. 7, we use $n^* \equiv \langle n(t^*)\rangle$ as a measure of z, since at t^* correlations are very pronounced and $\langle n(t)\rangle$ is nearly maximal [47]. Using the values of $S_{\rm conf}$ from Ref. [27], we find a linear relationship between n^* and $1/S_{\rm conf}$ (Fig. 7(a)),

$$n^* - 1 \propto \frac{1}{S_{\rm conf}}. \tag{8}$$

This finding is consistent with the possibility that $n^* - 1$ can be regarded as a measure of z and provides a quantitative connection between SHD clusters and the AG approach [48]. It is necessary to subtract one from n^* to obtain direct proportionality, implying that a cluster of unit size does not correspond to a CRR[30]. Equation (8) provides a clear link between a cluster property, n^*, and a property of the PEL, $S_{\rm conf}$. Since $S_{\rm conf}$ and the diffusion constant D are related [27], we expect to find

$$D \sim e^{-A(n^*-1)/T}. \tag{9}$$

Indeed, Fig. 7(b) confirms this expectation.

4. Transitions between Inherent Structures

It has been suggested [18,21] that diffusion in cold liquids is a consequence of a more subtle kind of heterogeneities. In principle, these heterogeneities occur when the system moves between consecutive local minima (IS) in the potential energy landscape (PEL). In this section we show that clusters between IS can be identified from simulations of water.

As Stillinger suggested [18,20,21], the PEL is expected to be very rough and contiguous basins may be grouped together to form wider 'craters' or 'megabasins'[49,50]. In this picture, the elementary transition processes (identified with the β-relaxation) connect contiguous basins and requires only local rearrangements of small number of particles. The escape from one deep basin within a large-scale 'crater' to another (α-relaxation) requires a lengthly directed sequence of elementary transitions and it will acquire a net elevation change many times that of the former. In this way, the diffusion of the molecules (which move forming SHD-clusters) are a consequence of the elementary changes between basins (or IS-transitions).

The trajectory of the point representing the system on a given liquid potential can be followed with molecular dynamics simulation, and an energy minimization algorithm can be implemented to find local minima in the PEL[17]. By means of this method, the motion in phase space is converted into a minimum-to-minimum trajectory, or *IS trajectory*. Here we study the IS trajectory at $T = 180$ K (below $T_{MCT} \approx 193$ K) using the SPC/E potential. The possibility of performing such a study below T_{MCT}, with a very fine time coarse graining, allows us to examine the structural changes that accompany the basin transitions and to describe an elementary step of the diffusive process.

Our system is started from equilibrated configurations at 190 K, which relax for nearly 920 ns at 180 K before we record and analyze the trajectory. At such a low temperature, a slow aging in the trajectory could be present, however, the aging should not affect the qualitative picture we present. We have generated one trajectory of 30 ns, sampling configurations at each 1 ps. For each configuration we find the corresponding IS using conjugate gradient minimization. In this way, we obtain 30,000 configurations with the corresponding IS. Since we could miss some IS transitions with 1 ps sampling, we also ran four independent 20 ps simulations sampling the IS at 4 fs. In this way, we obtain another 20,000 configurations with the corresponding IS.

Figure 8 shows an example of the potential energy of the IS, $E_{IS}(t)$, and the mean square displacement of the oxygen atoms $\langle r^2(t)\rangle$ starting from a single arbitrary starting time. At $T = 180$ K, the slowest collective relaxation time $\tau_\alpha > 200$ ns [4]. The IS trajectory in Fig. 8 has a mesh of 1 ps and covers a total time of 30 ns. In this time interval, $\langle r^2(t)\rangle$ is about $1\mathring{A}^2$, i.e., much less than the corresponding value of the average nearest neighbor distance of $2.8\mathring{A}$. Figure 9 shows an enlargement of the IS trajectory using a much smaller time mesh (4 fs, two times the simulation time step). Fig. 9 shows that changes occur via discrete transitions, with an average duration of ≈ 0.2 ps. The transitions are characterized by an energy change of $\approx 10-20$ kJ/mol and an oxygen atom square displacement of the order of 0.01 $\mathring{A}^2$; they appear to constitute the elementary step underlying the diffusional process in the system.

We note that it is impossible to identify the transition unless the quenching time step is of the order of the simulation time step. This is in distinct contrast to a Lennard-Jones liquid, where such small continuous changes are not present [22]. The difference, we will show, is attributable to the hydrogen bonds. This time scale for IS transitions is in accord with other work on the TIPS2 water model, at $T = 298$ K [51].

In Fig. 9, we see the sharp changes in $E_{IS}(t)$ coincide with the sharp changes in $\langle r^2(t)\rangle$. This confirms that the system is repeatedly visiting specific configurations, since $\langle r^2(t)\rangle$ and $E_{IS}(t)$ take on discrete values. The results shown in Fig. 9 imply that the system often returns to the original basin because of both the difference in energy and the displacement approach zero at the end of the time interval.

To aid in understanding the distribution of the displacements during the IS changes (such as those between the two IS labeled A and B in Fig. 9(b)), Fig. 10 shows the displacements u of all 216 individual molecules from IS-A to IS-B. We see that there is a relatively small set of molecules with a large displacement. Interestingly, we find that this set of molecules forms a cluster of bonded molecules. Indeed, for all cases studied, we found that the set of molecules which displace most form a cluster of bonded molecules.

As it was shown in [53], although a subset of "highly mobile" molecules is identifiable using a pre-defined threshold value in a single basin change, there is no unambiguous general criterion for identifying the molecules responsible for a single basin transition.

Water is characterized by a tetrahedral hydrogen bond (HB) network. However, many experiments suggest that this network has defects, such as an extra (fifth) molecule in the first coordination shell[54,55]. We note that the clusters identified in the IS-transition are related to the re-structuring of the hydrogen bond (HB) network that characterizes water. This can be observe in Fig. 11 where we show the number of molecules with a coordination number equal to 3 (low-density defect), 4 or 5 (high-density defect) as a function of time for a characteristic time interval and contrasts this data with the time dependence of $\langle r^2(t)\rangle$. A clear anti-correlation is observed between the time dependence of the number of 3 and 5-coordinated molecules compared to the time dependence of the 4-coordinated molecules. The changes in the HB are clearly connected with jumps in $\langle r^2(t)\rangle$ indicating that the HB-changes occurs when the system changes IS [53].

5. Discussion

We have shown that SHD are present in MD simulations of water. In accordance with experiments in colloids and simulations of simple liquids, the SHD-clusters grow in size as the temperature decreases. In connection with the Adam-Gibbs theory we have shown that the average mass of the SHD-clusters can be interpreted as the mass of the cooperative rearranging regions defined by Adam-Gibbs. By studying the PEL, we have shown that clusters of molecules with larger displacements are present in IS-transitions. A quantitative connection between clusters found with the SHD-analysis and those found in transitions between IS will be presented elsewhere. Preliminary calculations show that clusters between *consecutive* IS and those obtained between the corresponding configurations in the MD trajectory are uncorrelated. Such a correlation occurs for SHD-clusters and clusters between two IS sampled every approximately $t > 1$ ps.

REFERENCES

1. F. Sciortino, P. Gallo, P. Tartaglia, and S.-H. Chen, Phys. Rev. E **54**, 6331 (1996).
2. F. Sciortino, L. Fabbian, S.-H. Chen, and P. Tartaglia, Phys. Rev. E **56**, 5397 (1997).
3. W. Kob and H.C. Anderson, Phys. Rev. Lett **73**, 1376 (1994).
4. F. W. Starr, F. Sciortino, and H.E. Stanley, Phys. Rev. E **60**, 6757 (1999).
5. K. Schmidt-Rohr and H.W. Spiess, Phys. Rev. Lett **66**, 3020 (1991).
6. J. Liesen, K. Schmidt-Rohr, and H.W. Spiess, J. Non-Cryst. Solids **172-174**, 737 (1994).
7. A. Heuer, M. Wilhelm, H. Zimmermann, and H.W. Spiess, Phys. Rev. Lett **75**, 2851 (1995).
8. M.T. Cicerone and M.D. Ediger, J. Chem. Phys. **103**, 5684 (1995).
9. F. Fujara, B. Geil, H. Sillescu, and G. Fleischer, Z. Phys. B **88**, 195 (1992).
10. M.T. Cicerone, F.R. Blackburn, and M.D. Ediger, J. Chem. Phys. **102**, 471 (1995).
11. M. Hurley and P. Harrowell, Phys. Rev. E **52**, 1694 (1995).
12. A.I. Mel'cuk, R.A. Ramos, H. Gould, W. Klein, and R.D. Mountain, Phys. Rev. Lett **75**, 2522 (1995).
13. T. Muranaka and Y. Hiwatari, Phys. Rev. E **51**, R2735 (1995).
14. W. Kob, C. Donati, S.J. Plimpton, P.H. Poole, and S.C. Glotzer, Phys. Rev. Lett. **79**, 2827 (1997).
15. P.H. Poole, S.C. Glotzer, A. Coniglio, and N. Jan, Phys. Rev. Lett. **78**, 3394 (1997).
16. M. Goldstein, J. Chem. Phys. **51**, 3278 (1969).
17. F. H. Stillinger and T. A. Weber, Phys. Rev. A **28**, 2408 (1983).
18. F. H. Stillinger, Science **267**, 1935 (1995).
19. C.A. Angell, Science **267**, 1924 (1995).
20. S. Sastry, P.G. Debenedetti, and F.H. Stillinger, Nature (London) **393**, 554 (1998).
21. P.G. Debenedetti, and F.H. Stillinger, Nature (London) **410**, 259 (2001).
22. T.B. Schrøder, S. Sastri, J.C. Dyre, and S.C Glotzer, J. Chem. Phys. **112**, 9834 (2000).
23. A. Heuer, Phys. Rev. Lett. **78**, 4051 (1997); S. Buchner and A. Heuer, Phys. Rev. E **60**, 6507 (1999).
24. L. Angelani, G. Parisi, G. Ruocco, and G. Villani, Phys. Rev. Lett. **81**, 4648 (1998).
25. F. Sciortino, W. Kob, and P. Tartaglia, Phys. Rev. Lett. **83**, 3214 (1999).

26. G. Adam and J. H. Gibbs, J. Chem. Phys. **43**, 139 (1965).
27. A. Scala, F. W. Starr, E. La Nave, F. Sciortino and H. E. Stanley, Nature (London) **406**, 166 (2000).
28. W. Kob, C. Donati, S. J. Plimpton, P. H. Poole, and S. C. Glotzer, Phys. Rev. Lett. **79**, 2827 (1997).
29. C. Donati J. F. Douglas, W. Kob, S. J. Plimpton, P. H. Poole, and S. C. Glotzer, Phys. Rev. Lett. **80**, 2338 (1998).
30. S. C. Glotzer, P. H. Poole, W. Kob, S. J. Plimpton, Phys. Rev. E **60**, 3107 (1999).
31. M. Hurley and P. Harrowell, Phys. Rev. E **52**, 1694 (1995).
32. B. Doliwa and A. Heuer, Phys. Rev. Lett. **80**, 4915 (1998).
33. K. Schmidt-Rohr and H. W. Spiess, Phys. Rev. Lett **66**, 3020 (1991); R. Böhmer et al., Europhys. Lett. **36**, 55 (1996); B. Schiener et al., Science **274**, 752 (1996); W. K. Kegel, and A. van Blaaderen, Science **287**, 290 (2000).
34. M. T. Cicerone and M. D. Ediger, J. Chem. Phys. **103**, 5684 (1995).
35. H. Sillescu, J. Non-Cryst. Solids **243**, 81 (1999); M.D. Ediger, Ann. Rev. Phys. Chem. **51**, 99 (2000).
36. E. R. Weeks et al., Science **287**, 627 (2000).
37. H. J. Berendsen et al. J. Phys. Chem. **91**, 6269 (1987).
38. J. P. Hansen and I. R. McDonald, *Theory of Simple Liquids* (Academic Press, London, 1986).
39. Y. Gebremichael T. B. Schrøder, F. W. Starr, and S. C. Glotzer, Phys. Rev. E **64**, 051503 (2001).
40. Alternatively, we also consider using a separation of 0.35 nm, the distance criterion commonly used by hydrogen bond definitions [F. Sciortino and S. L. Fornili, J. Chem. Phys. **90**, 2786 (1989)]. Preliminary calculations indicated this alternative choice does not qualitatively affect our results.
41. C. Bennemann, J. Baschnagel, and S. C. Glotzer, Nature **399**, 246 (1999).
42. Y. Gebremichael, T. B. Schrøder, and S. C. Glotzer, Abstr. Am. Chem. Soc. **220**, 412 (2000).
43. For a clear description of the physical basis of AG theory, see P.G. Debenedetti, *Metastable Liquids* (Princeton University Press, Princeton, 1996).
44. F. H. Stillinger, and T. A. Weber, J. Phys. Chem. **87**, 2833 (1983); F. Sciortino et al., Phys. Rev. Lett. **83**, 3214 (1999); B. Coluzzi, G. Parisi, and P. Verrocchio *ibid.* **84** , 306 (1999).
45. S. Sastry, Nature (London) **409**, 164 (2001); S. Mossa et al., Phys. Rev. E **65**, 041205 (2002).
46. J. Dudowicz, K. F. Freed, and J. F. Douglas, J. Chem. Phys. **111**, 7116 (1999).
47. The maximum of $\langle n(\Delta t)\rangle$ occurs at time slightly before t^*. Our conclusions are unaffected by choosing n^* or the maximum of $\langle n(\Delta t)\rangle$.
48. This connection relies on the assumption that the T dependence of $s_{\mathrm{conf}}(z)$ is weak in comparison to that of S_{conf}, as can be expected since $z \ll N$ and the configurational entropy is an extensive property.
49. B. Doliwa and A. Heuer, Phys. Rev. E **67**, 030501(R) (2003); Phys. Rev. E **67**, 031506 (2003).
50. R.A. Denny, D.R. Reichman, and J.-P. Bouchaud, Phys. Rev. Lett. **90**, 025503 (2003).

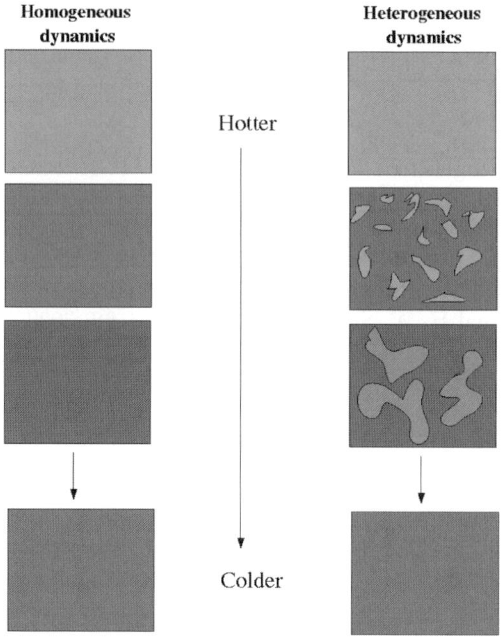

Figure 1. Two possible scenarios proposed to describe diffusion in cold liquids. In the homogeneous scenario molecules relax in the same way, while in the heterogeneous scenario, sets of more mobile molecules (in comparison to the average motion of the molecules in the system) form patches or clusters. The size of these clusters increases upon cooling.

51. I. Ohmine, and H. Tanaka, Chem. Rev. **93**, 2545 (1993) and references therein.
52. M. Matsumoto and I. Ohmine, J. Chem. Phys. **104**, 2705 (1996); I. Ohmine and S. Saito, Acc. Chem. Res. **32**, 741 (1999), and references therein.
53. N. Giovambattista, F.W. Starr, F. Sciortino, S.V. Buldyrev, and H.E. Stanley, Phys. Rev. E **65**, 041502 (2002).
54. *Hydrogen Bonded Liquids*, edited by J. Dore and J. Texeira (Kluwer, Dordrecht, 1991), pp.171-183.
55. E. Grunwald, J. Am. Chem. Soc. **108**, 5719 (1986); A.H. Narten and H.A. Levy, Science **165**, 447 (1969); P.A. Giguere, J. Chem. Phys. **87**, 4835 (1987); G.E. Walrafen, M.S. Hokmabadi, W.H. Yang, Y.C. Chu, and B. Monosmith, J. Phys. Chem. **93**, 2909 (1989).

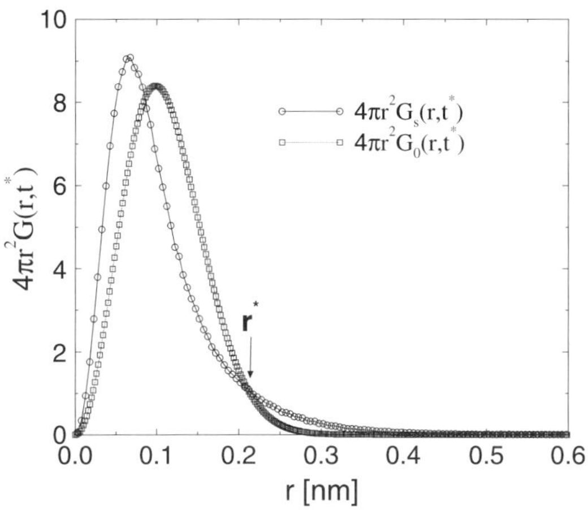

Figure 2. Van Hove correlation function $G_s(r, t^*)$ and its Gaussian approximation $G_0(r, t^*)$ obtained using $\langle r^2(t^*) \rangle$, for $T = 220$ K. The tails of the distributions cross at $r^* \approx 0.225$ for all temperature.

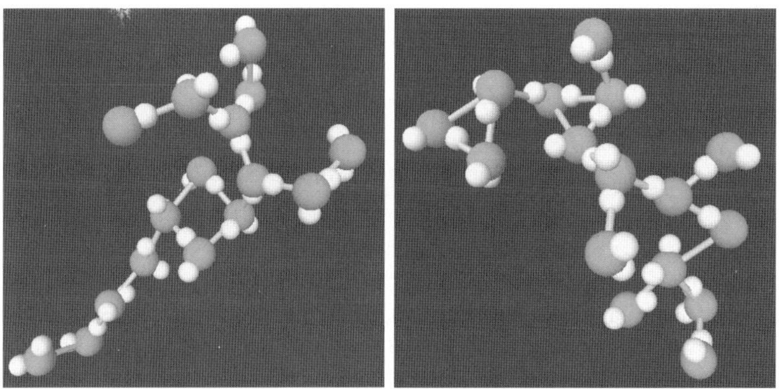

Figure 3. Two of the larger clusters of mobile molecules found at $T = 260$ K defined with an observation time $\Delta t = t^* \approx 3$ ps. Tubes connect neighboring molecules whose oxygen-oxygen distance is less than 0.315 nm, the first minimum in the oxygen-oxygen radial distribution function.

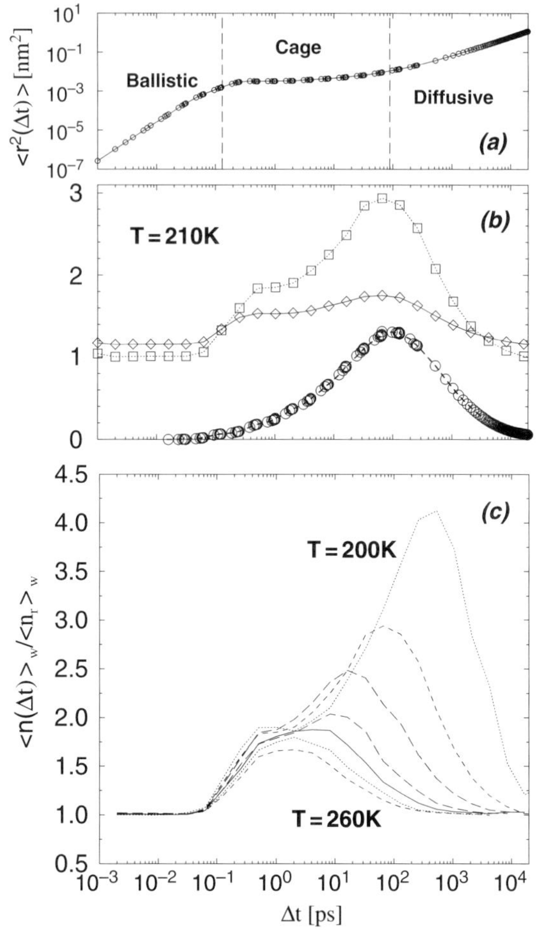

Figure 4. (a) Mean square displacement $\langle r^2(\Delta t)\rangle$ at $T = 210$ K showing the ballistic, cage and diffusive regimes (separated by dotted lines). (b) Average number of molecules $\langle n(\Delta t)\rangle$ (diamonds) and normalized weight cluster size $\langle n(\Delta t)\rangle_w$ (squares). The behavior of all quantities correlate with $\langle r^2(\Delta t)\rangle$. The maxima of $\langle n(\Delta t)\rangle_w$ and $\langle n(\Delta t)\rangle$ occur at times slightly smaller than the time for the maximum in $\alpha_2(\Delta t)$ (circles), the non-Gaussian parameter. (c) Weight cluster size $\langle n(\Delta t)\rangle_w / \langle n_r\rangle_w$ for temperatures ranging from 200 K to 260 K in intervals of 10 K. Note the T-independent plateau at the crossover from ballistic motion.

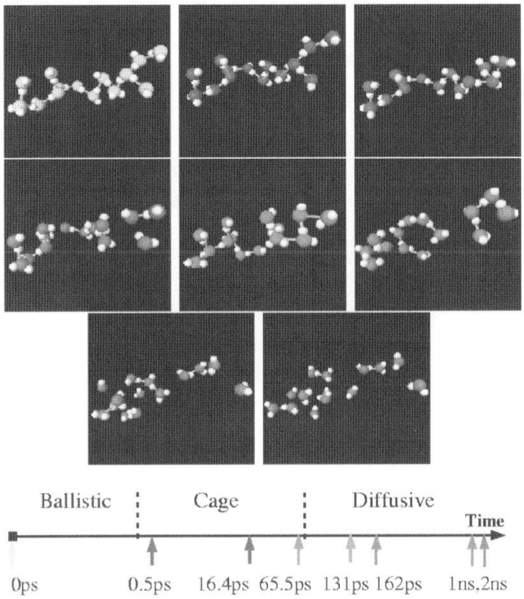

Figure 5. Time-evolution of a cluster of mobile molecules identified with $\Delta t = t^* \approx 65$ ps at $T = 210$ K. Time increases form left-to-right and top-to-bottom. Times are indicated along the time-axis at the bottom of the figure. The three regimes (ballistic, cage and diffusive) defined in Fig. 4(a) are also indicated along the time-axis.

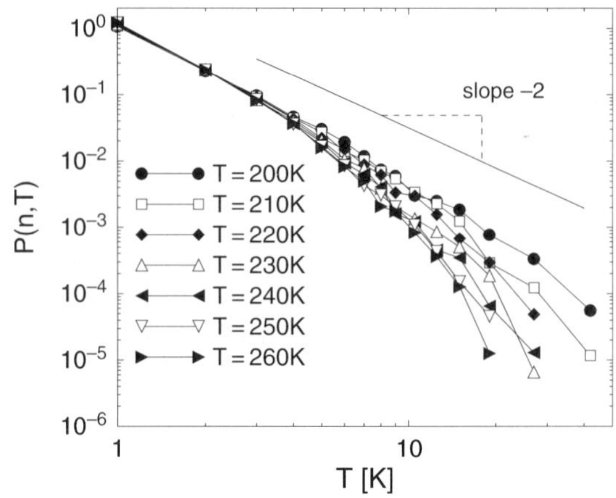

Figure 6. Probability distribution $P(n,T)$ for the number of molecules n in clusters (log-log scale) at temperature T. Data can be well fitted by $P(n,T) \sim n^{-\tau(T)} \exp(-n/n_0(T))$. Values of the parameters in this eq. are in TableI.

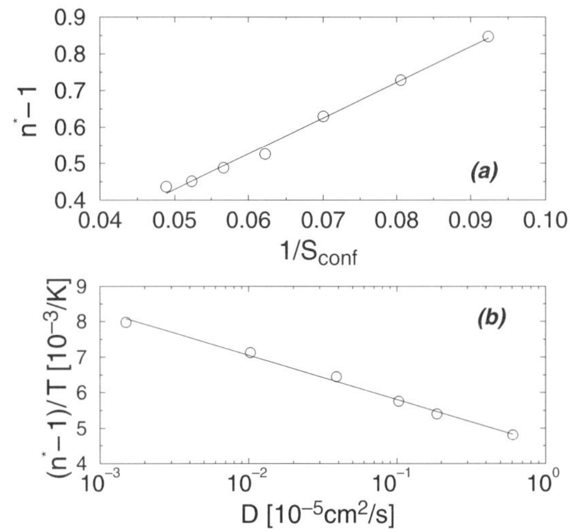

Figure 7. (a) The average cluster size n^* is proportional to the inverse of the configurational entropy $S_{\rm conf}$ suggesting that $n^* - 1$ can be used as a measure of the size of the cooperatively rearranging regions hypothesized by Adam and Gibbs. (b) Log-linear plot of $(n^* - 1)/T$ as a function of the diffusion constant D. The AG prediction $D \sim \exp(A/TS_{\rm conf})$ implies that $\log D \sim (n^* - 1)/T$. This relationship holds for almost three decades in D.

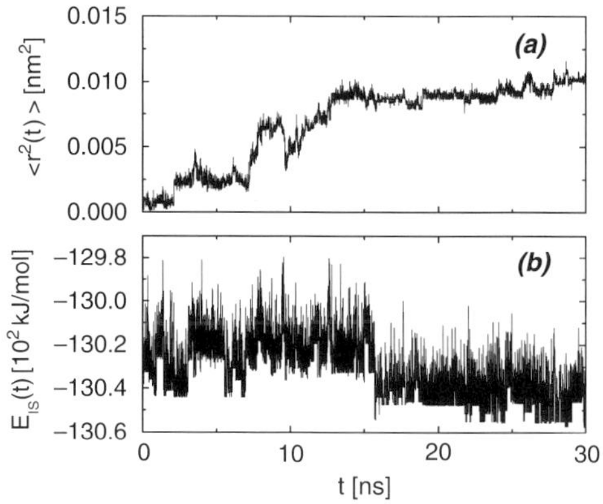

Figure 8. (a) Mean square displacement and (b) IS energy for the inherent structures as a function of time for the studied 216-molecule system. The time interval between adjacent IS in both figures is 1 ps. While it is possible to track IS transitions from the potential energy, it is not the case for the mean square displacement. Note the amplitude of the peaks of the potential energy is $\approx 10 - 20$ kJ/mol, the same order of magnitude as hydrogen bond energy.

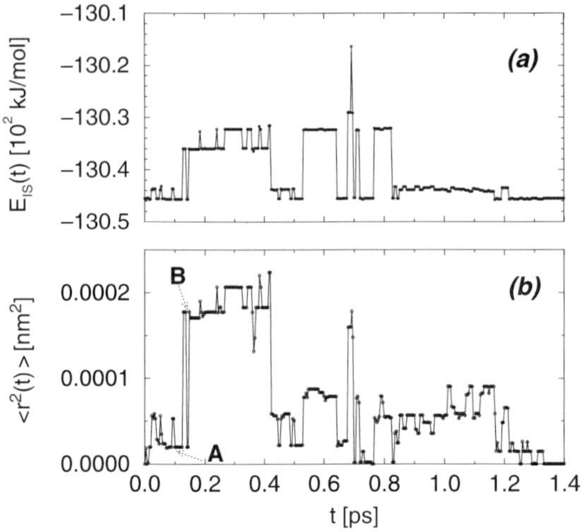

Figure 9. (a) IS energy and (b) mean square displacement for the IS obtained using a sampling interval of 4 fs, two times the simulation time step. The correlation between E_{IS} and $\langle r^2(t) \rangle$ is evident. Also, we see that it is necessary to sample IS with a mesh of the order of the simulation time step to detect all the IS visited by the system.

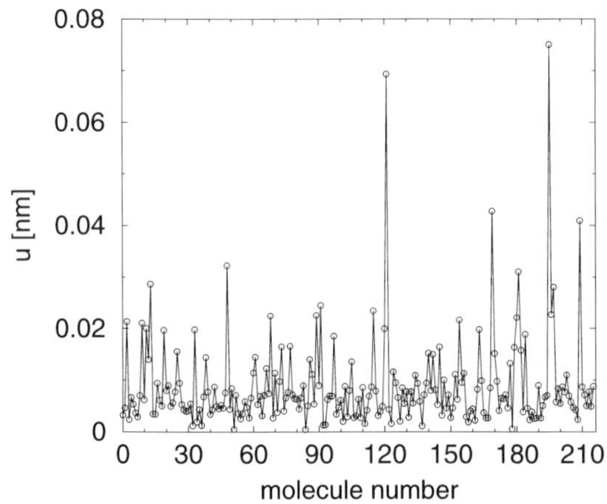

Figure 10. Displacement of each molecule in the transition from IS-A to IS-B shown in Fig. 9(b).

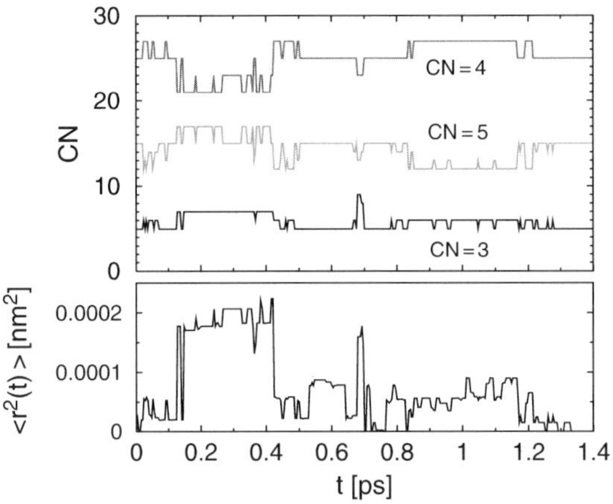

Figure 11. Number of molecules with coordination number CN equals 3,4 and 5 versus time, for the IS corresponding to Fig. 9. The plot for $CN = 4$ is shifted down by 400 units for better comparison. Also shown is $\langle r^2(t) \rangle$. We see how the tetrahedral network acquires both types of defects (CN equals to 3 and 5) while the system explores different IS.

Unifying Concepts in Granular Media and Glasses
A. Coniglio, A. Fierro, H.J. Herrmann and M. Nicodemi (editors)

Glass States in Dense Attractive Micellar Systems

F. Mallamace[ab], M. Broccio[a], W.R. Chen [b], A. Faraone[ab], and S.H. Chen[b]

[a]Dipartimento di Fisica and INFM, Universita' di Messina, Italy.

[b]Department of Nuclear Engineering, Massachusetts Institute of Technolgy, Cambridge, USA.

Recent mode coupling theory (MCT) studies show that if a short-range attractive interaction is added to the pure hard sphere system, one may observe a new type of glass. This in addition to the glass-forming mechanism due to cage effects in the hard sphere system. Furthermore, within a certain volume fraction range where the two glass-forming mechanisms nearly balance each other, varying the external control parameter, the effective temperature, makes the glass-to-liquid-to-glass re-entrance and the glass-to-glass transitions possible. Here, by using small angle neutron scattering and photon correlation measurements in a micellar system, we present evidence on this complex phase behavior. Finally, in agreement with MCT, we found an end point on the glass-to-glass transitions line beyond which the two glasses become identical in their local structure and the long time dynamics.

1. Introduction

Colloids and soft-materials are of large interest in science and technology [1]. These disordered systems can evolve under appropriate conditions (density, temperature and load) in a jammed status[2] where the system particle is caught in a small region of phase space with no possibility of escape. In these conditions materials resemble solids because they are driven in a structural arrest (SA) status by thermodynamical constraints or an externally applied stress. But, on the contrary of ordinary solids, jammed systems can easily break up if the direction of the applied stress changes even by a small amount. A canonical example is a pile of sand, which appears as solid: the upper surface slopes and sustains its shape despite gravity; but if one tilts or vibrates the pile, the grains shift and the solid melts. These unusual mechanical properties of fragile matter require a new description and new theoretical models and applications have been recently proposed [3,4].

Originally jamming has been studied in macroscopic (from micrometer to millimeters scales) systems with a dynamics driven by surface friction and inter-particle repulsion[3,4]. In that case, particles are constrained through an applied stress and individual particles are large, so that there is no thermal motion. Recently, by considering colloids with attractive interactions, these constrains have been removed. On this basis, a "general" phase diagram has been proposed for which jamming transitions can depend on density,

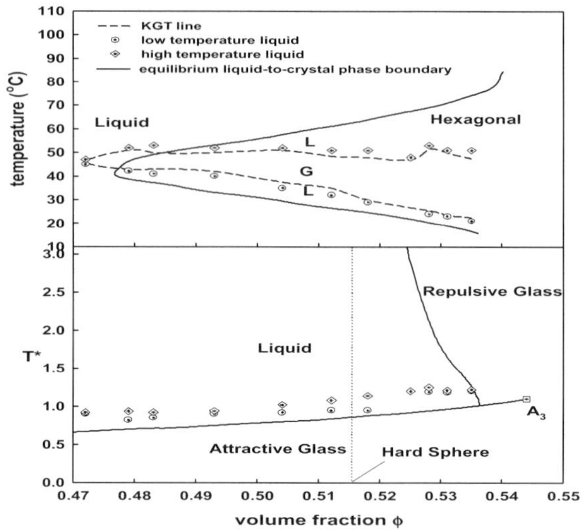

Figure 1. (bottom panel) The MCT phase diagram for an adhesive hard sphere system ($\epsilon = 0.03$). Phase diagram of the $L64/D_2O$ system (top panel). The dotted line marks the equilibrium phase boundary between the disordered micellar phase (phase) and the ordered, metastable, liquid crystal hexagonal phase. Solid symbols represent L(liquid)-G(glass)-L(liquid) transition boundaries.

load or stress and temperature. Such a phase diagram was proposed with the idea that apparently different materials like granular matter, foams, gels, glasses, colloids have common features and, therefore, their physical properties may be classifed under the same unique rubric. From this description, some interesting questions emerge: a glass may have a lower glass transition temperature under high shear stress, and transitions from ergodic to non-ergodic states (e.g. sol-gel, liquid-glass transitions and the related aging processes) may have a common description. Likewise, if temperature and applied stress play similar roles, it is possible that macroscopic athermal jamming system, like granular materials or foams, may be described in terms of an effective temperature (T^*).

The addressing of these new conceptual problems related to attractive systems involves theory, simulation and experiments and represents today one of the most important and challenging subjects of statistical physics. New and known approaches have been used [3,5–8]. For example frustrated percolation describes similarities between granular jamming and glass transition [5], whereas several mode coupling theory (MCT) approacheses confirm that the structural arrest (SA) in attractive colloids have two order parameters, i.e. temperature and density [7,8]. To explore these emerging situations, different experimental techniques like viscoelasticity[9,10], photon-correlation spectroscopy (PCS)[11–15],

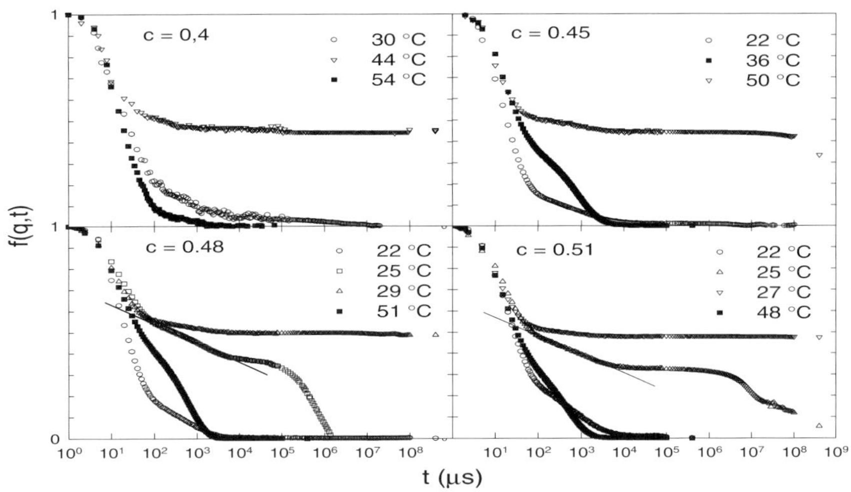

Figure 2. Some ISF measured at various concentrations and temperatures. In the intermediate temporal region as proposed by the MCT the ISFs are characterized by a logarithmic decay (long dashed lines).

X-ray scattering[16], neutron scattering[17] and imaging [18] have been used.

Important suggestions come from the application of MCT to colloidal-like attractive systems[19]. On the contrary of the known MCT results on hard sphere where the kinetic glass transition (KGT), with concentration as the only order parameter, is explained in terms of cage effects, the interparticle attractions (dominated by clustering processes) give rise to a new and rich physical scenario. In this latter case, all dynamical processes are due to the competition between cage (concentration or volume fraction) and clustering (temperature-attraction) effects [7,8]. According to the original speculation on jamming, the system phase diagram presents ergodic non-ergodic transitions having concentration and temperature as order parameters. More precisely, the attractive hard sphere systems (AHS) the phase diagram have: (i) an inverted binodal curve (with a critical point) at moderated concentration; (ii) two kinetic glass transition lines, one corresponding to the glass line of hard sphere systems at high composition and the other extending to much lower concentrations. The former is attributed (with the usual mechanism due to excluded volume) to the repulsive part of the interaction, the latter (temperature dependent) to the attractive part of the potential; (iii) the attractive branch crosses the repulsive one with an end point of type A_3 (a higher order glass transition and a glass-glass transition line); (iv) a cusp-like singularity with a glass-liquid-glass reentrant behavior. In particular, a new phenomenon arises: at the glass-glass transition, due to the crossing of the two glass lines, the density decays are characterized by a logarithmic time dependence[20].

It must be noticed that some attractive colloidal systems (microemulsions and micellae) present in their phase diagram a well defined percolation (or gelation) line (PT)[21].

This $c - T$ dependent line of transition from an ergodic to a non-ergodic state cutting across the phase diagram starting near the critical point to at very high c. In the past [22,23], the observation of the onset of a metastable glass state due to the attraction was attributed to a static percolation process, but several results in copolymer micellar systems give definitive indication that the PT and the attractive glassing are distinct dynamic transitions [24,21].

MCT describes the phase behavior of the AHS by the volume fraction of the disperse phase ϕ and the fractional attractive well width $\epsilon = (R - R')/R$ ($R' = 2a$, the diameter of a repulsive hard core and R the diameter of the particle). For a given ϵ, aside from ϕ, a second external control parameter, the effective temperature $T^* = k_B T/u$, is introduced and the loss of ergodicity can take place by either increasing ϕ or changing T^*(**Fig.1**, bottom panel, shows the MCT calculated AHS phase diagram for $\epsilon = 0.03$). Hence, by varying ϕ and T^*, it is possible to explore the very rich physics of attractive colloids. Several experimental studies have confirmed some of the MCT predictions, such as the reentrant glass-to-liquid-to-glass transitions[8,7] and the logarithmic relaxation of the glass dynamics in the liquid state in the vicinity of the A_3 point; these studies have been mainly conducted by using photon correlation spectroscopy (PCS) in polymer grafted colloids characterized by a polymer-induced depletion attraction [15] and in a triblock copolymer micellar systems[11,17].

Motivated by our previous experience in the study of percolation and KGT processes, here we report a large series of small angle neutron scattering (SANS) and PCS experiments performed in a micellar solution in heavy water, which exposes all the characteristics of the AHS[24,17]. We show that a combination of these two techniques can successfully confirm, in a quantitative way, all the MCT suggestions for AHS systems: the logarithmic decays of the density correlation near the attractive KGT branch, the predicted reentrant glass-to-liquid-to-glass phenomenon, the glass-to glass transition, and most importantly, the existence of the A_3 singularity. Furthermore, we show that the use of SANS intensity distributions in absolute unities and a novel scaling approach in describing the scattered intensities, give the possibility to map out the complete structural arrest phase boundaries.

2. The system

The micellar system we study is described in detail in ref.[24]. It is an aqueous solution of a non-ionic triblock copolymer made of polyethylene oxide (PEO) and polypropylene oxide (PPO), whose chemical structure is $[PEO_{13} - PPO_{30} - PEO_{13}]$ ($L64$-Pluronic, for short). The main property of the copolymer molecules is that in water they are T-dependent surfactants that self-assemble forming hydrated monodisperse spherical micelles in a wide $T - c$ range. At low T, both PEO and PPO are hydrophilic, so that $L64$ chains readily dissolve in water, existing as unimers. As the temperature increases there is a decrease in the hydrogen bond formation between water and polymer molecules, PPO tends to become less hydrophilic faster than PEO, and the copolymer acquires surfactant properties and aggregates to form micelles. Since at higher temperatures water becomes progressively a poor solvent to both PEO and PPO chains, the effective micellar interaction becomes attractive. The evidence for the increasing short-range micellar

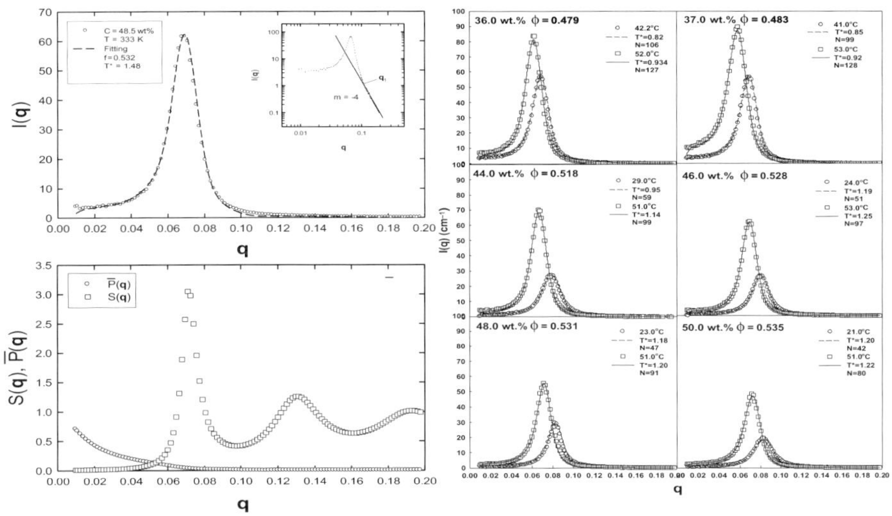

Figure 3. **(3a)** (Two panels on the left) The model fitting for a SANS intensity distribution in absolute scale obtained from a 48.5% micellar solution at $T = 333K$ (top panel) In the inset the same data are plotted in a log-log scale (for large q, the SANS intensity decreases as q^{-4}, in agreement with the Porod's law). **(bottom panel)** The normalized intra-particle structure factor $\overline{P}(q)$ (circles) and the inter-particles structure factor $S(Q)$ (squares) for the case reported in the top panel. **(3b)** (Six panels on the right) The theoretical fits of SANS intensity distributions for the studied micellar system at different concentrations and temperatures.

attraction as a function of T comes from the existence of a lower consolute critical point ($c = 5\ wt\%$, $T_c = 330.32\ K$) and a $T - c$ dependent percolation line cutting across the phase diagram[24,11]. The system phase diagram for high concentration, shown in the **Fig. 1** (top panel) also contains the equilibrium phase boundary (liquid-to-crystal) between the disordered micellar phase and the ordered liquid crystalline (hexagonal) phase and the equilibrium crystal-to-crystal phase boundary [25]. The same figure also reports data points of a re-entrant liquid-glass-liquid, as predicted by MCT[7] and determined by scattering[17].

We have explored the dynamically arrested states and their structure in this micellar system, as a function of T at different c (or ϕ). The PCS were made at a scattering angle $\theta = 90°$, using a continuous wave solid state laser operating at $50mW$ ($\lambda = 5120$Å). The intensity data were also corrected for turbidity and multiple scattering effects. PCS data have been taken using a digital full-correlator with a logarithmic sampling time scale which allows an accurate description of both the short and the long time regions of the

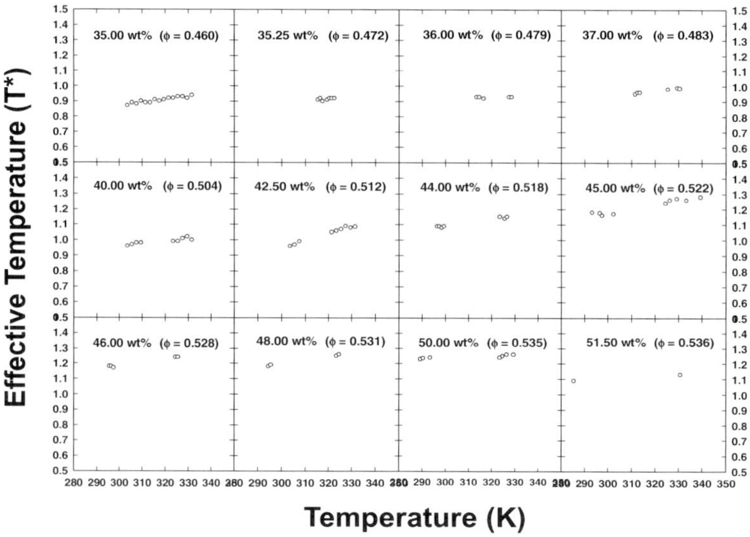

Figure 4. The variations of the fitted effective temperature T^*, as a function of the temperature T for several volume fractions.

intermediate scattering function (ISF), from 1μsec up to 100 sec. Since the system shows nonergodic behaviors, particular care has been taken for proper intensity averages[11].

SANS measurements were performed at $NG7$, $47m$ SANS spectrometer in the NIST center for Neutron Scattering Research, and a SAND station at the Intense Pulsed Neutron Source (IPNS) in Argonne National Laboratory. At $NG7$, incident monochromatic neutrons of wave length $\lambda = 5$Å with $\Delta\lambda/\lambda = 10\%$ was used. Sample to detector distance was fixed to cover a wave vector transfer (q) range from 0.008Å^{-1} to 0.3Å^{-1}. At IPNS a pulse of withe neutrons was selected with an effective wave length range from 1.5Å to 14Å. In SAND the reliable q-range covered in the measurements were from 0.004Å^{-1} to 0.6Å^{-1}; q-resolution functions of both these SANS spectrometers are Gaussian. It is essential that we apply these resolution broadening to the theoretical cross-section when fitting the intensity data. The measured intensity was properly corrected (background, empty cell contributions) and normalized by a reference scattering intensity).

3. Experiments

3.1. Photon Correlation Spectroscopy

Following the method used for colloidal hard sphere glasses [26], by means of PCS we measure the time intensity correlation function,$g^{(2)}(q,t)$, directly related to the interparticle dynamical structure factor $S(q,t) = \langle \delta\rho(q,o)\delta\rho^*(q,t)\rangle$ where $\delta\rho(q,t)$ is the spatial Fourier component of concentration fluctuations. The intermediate scattering functions (ISF), **Fig. 2**, defined through of the dynamical structure factor, $S(q,t)$,

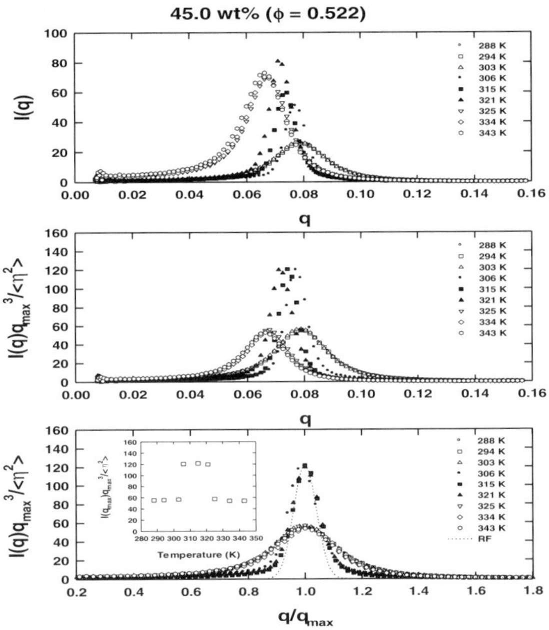

Figure 5. A series of SANS intensity distributions and their associated scaling plots for the $c = 45wt\%$ sample, at $\phi = 0.522$, at different temperatures. The inset gives the peak height of the scaling plots as a function of T. The variation of the peak heights,as a function of T, reported in the inset indicates the re-entrant liquid-to-attractive glass-to-liquid transition as temperature increases.

as $f(q,t) = S(q,t)/S(q,0)$ are then obtained straightforwardly using Siegert relation $f(q,t) = \left[(g^{(2)} - 1)/c\right]^{1/2}$. The measurements are usually taken at constant c and increasing T. We observe in the measured ISFs (**Fig. 2**) a progressive slowing down of their decay, up to a point where it becomes flat, indicating the structural arrest. The SA is experimentally identified by the breakdown of Siegert relation. The SA will appear only above that threshold; below it, one observes the usual stretched exponential behavior of dense colloids but no nonergodicity set in. In **Fig.2a** is evident in the intermediate temporal region of the ISF the MCT proposed logarithmic decay (long dashed lines).

The MCT predicts, for the ISF, the existence of a critical temperature T_c where the ergodic to nonergodic transition takes place and $f_q(t)$ tends to the finite plateau (the non ergodicity parameter, or Debye-Waller factor (DWF)) $f_q(t \to \infty,) = f_q^c > 0$. The separation parameter is $\sigma \approx (T_c - T)/T_c$. Three density relaxation regimes, with different temporal scales, are proposed. A short time region $t < t_0$ dominated by microscopic motions is followed by the β relaxation region that satisfies the scaling relation

$$f_q(t) = f_q^c + h_q G(t) = f_q + c_\sigma h_q g_\pm(t/t_\sigma) \tag{1}$$

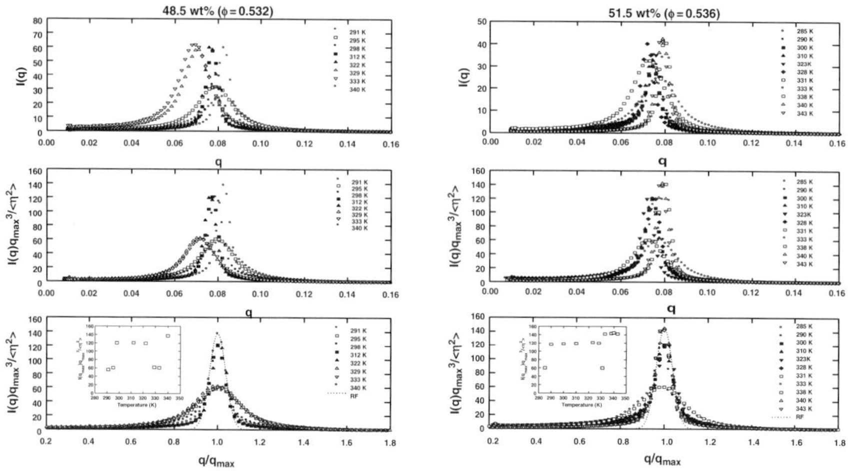

Figure 6. The SANS intensity distributions and their scaling plots for samples at $c = 48.5wt\%$ ($\phi = 0.532$) and $c = 51.5wt\%$ ($\phi = 0.536$) at a series of temperatures ranging from $291K$ to $340K$. In addition to all the similarities with **Fig. 5**, when T increases to $340K$, the system is driven into another glassy state peaked at $q = 0.082\text{Å}^{-1}$(i.e. by varying T, the system shows a liquid-to-glass-to-liquid-to-glass transition). The difference between these two glass states is in their degree of disorder.

where $G(t)$, the so-called β correlator, depends singularly on the time and the control parameter ($t_\sigma = t_0 |\sigma|^{-1/2a}$ and $c_\sigma = |\sigma|^{1/2}$) and h_q is the critical amplitude. The initial part ($g_\pm(t << t_\sigma) = (t/t_\sigma)^{-a}$), dominated by t_σ, gives the approach to f_q, while the final one follows von Schweidler's law $g_-(t_\sigma << t << t'_\sigma) = -h_q(t/t'_\sigma)^b$, with $t'_\sigma = t_0 |\sigma|^\gamma$ as the second time scale ($\gamma = (1/2a) + (1/2b)$). a ($0 < a < 0.5$) and b ($0 < b < 1$) exponents determined solely by the so-called exponent parameter λ as $\lambda = \Gamma^2(1-a)/\Gamma(1-2a) = \Gamma^2(1+b)/\Gamma(1+2b)$ where Γ is the Euler gamma function and λ is in turn determined by the static structure factor $S(q)$. Finally, for $t > t'_\sigma$, there is the α relaxation regime: $f_q(t) = f_q^c G(t/t'_\sigma)$. For hard spheres MCT gives $a = 0.301$ and $b = 0.545$[19,26].

Analyses of the reported experimental data give the parameter $b = 0.6$ in the β regime and the exponent parameter $\lambda = 0.7$, and the universal plot of von Schweidler law gives the exponent $\gamma = 2.3$ [11]. The emerging picture of these data was that, for the studied micellar system, the attractive contribution to the potential is not only the origin of the phase separation, but also that of the formation of ramified clusters, responsible of the percolation process and of the attractive glass branch. Here, starting from these results, we largely used the PCS technique in order to reveal a complete signature of the complex phase behavior of attractive systems suggested by MCT.

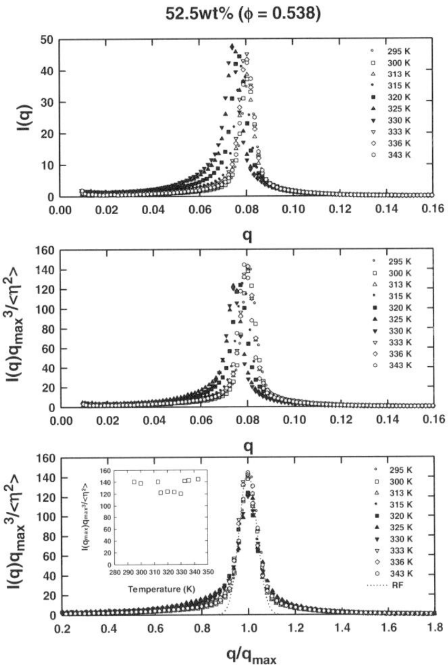

Figure 7. The SANS intensity distributions taken at $\phi = 0.538$. As it can be observed the much broader peak (liquid state) disappear and the T variations of the peak heights in the scaling plots given in the inset show a re-entrant repulsive glass-to-attractive-glass-to repulsive-glass transition.

3.2. Small Angle Neutron Scattering (absolute scale)

For a mono-dispersed macromolecular system the absolute scattered neutron intensity (in unit of cm^{-1}) is: [17]:

$$I(q) = cN \left[\sum_i b_i - \rho_w \nu_p \right]^2 \left| \overline{F}(q) \right|^2 S(q) \tag{2}$$

where $\left|\overline{F}(q)\right|^2$ is the normalized particle form factor and $S(q)$ the structure factor ($\overline{P}(q) = \left|\overline{F}(q)\right|^2$ is the so called normalized intra-particle structure factor), c the concentration of polymer (number of polymer/cm^3), N the aggregation number of polymers in a micelle, $\sum_i b_i$ sum of the coherent scattering lengths of polymer atoms, ρ_w the scattering length density of D_20 molecules and ν_p the molecular volume of the polymer.

Whereas $S(q)$ is calculated by using precise methods of statistical mechanics, for $\overline{P(q)}$ it is very important to consider the precise structure of the scattering object . For the calculation of such a quantity in a suspension of hydrated spherical micelles (like the system $L64/D_20$) we have developed a model that takes into account the polymer surfactant

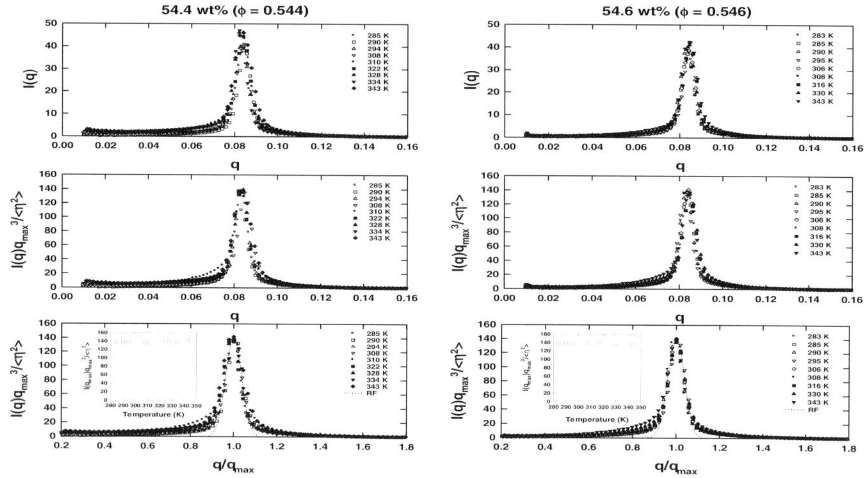

Figure 8. **(8a)** The SANS intensity data and its associated scaling distributions at $\phi(A_3) = 0.544$ (the volume fraction of the A_3 end point). All the scaled intensity curves collapse into one single master curve, indicating the fact that the first diffraction peak of $S(q)$ for all states are identical. The inset shows that all the scaling peaks have the identical height (~ 140), indicating two glasses with the same degree of local order. **(8b)**, The SANS intensity data and its associated scaling distributions at $\phi = 0.546$; as can be observed all the scaled intensities are again characterized by a unique T-dependent length scale and collapse into one single master curve showing an identical local structure of the two glasses.

properties [27]. In this model, named the cap-and-gown model, we have assumed that micelle has a compact spherical hydrophobic core of radius a, consisting of all the PPO segments with a polymer volume fraction in the core $\phi_p = 1$ (i.e. a dry core) and a diffuse corona region consisting of PEO segments and the solvent molecules. The model assumes that the radial distribution profile in the corona is a power-law decay in which the power n is determined by two geometrical constraints connecting it with the core radius. These constraints are: (i) the total volume of the polymer segments in the core is given by $(4\pi a^3/3) = N\nu_{PPO}$ where ν_{PPO} is the PPO segmental volume ($\nu_{PPO} = 95.4 \times 30$ Å^3). Similarly, the total volume of polymer segments outside the core is given by N and the volume integral of the PEO segmental distribution ν_{PEO} as $4\pi \int_a^\infty (a/r)^n r^2 dr = N\nu_{PEO}$ ($\nu_{PEO} = 72.4 \times 26$ Å^3). $\overline{P}(q)$ calculated in the frame of this model is only function of N.[17].

The micellar structure factor $S(q)$ was calculated in the same way used in the MCT approach to study the phase diagram of attractive colloids. We have worked with the same square well potential (hard sphere with an adhesive surface layer) defined in terms of T^*, ϵ and R . Then, we have solved (analytically to the first order in a series of small ϵ) the Ornestin-Zernike equation in the Percus-Yevick approximation obtaining a formula that give $S(q)$. The main point in this calculation for $I(q)$ was that an absolute SANS intensity distribution can be fitted uniquely with four parameters: N, ϕ , ϵ and the effective temperature T^*[17]. It should be noticed that the real particle diameter R is tied uniquely to the aggregation number N and the volume fraction ϕ through the relation $\phi = c\pi R^3/6N$.

Fig.3a (top panel) illustrates the model fitting, a SANS intensity distribution obtained from a 48.5% micellar solution at $T = 333K$ in absolute scale. The same data plotted in a log-log scale are given in the inset. It can be seen that for large q, the SANS intensity decreases as q^{-4}, in agreement with the Porod's law ($I(q) \sim Aq^{-4}$) as expected in a two-phase system with a sharp interface. **Fig.3a** (bottom panel) gives the normalized intra-particle structure factor $\overline{P}(q)$ (circles) and the inter-particles structure factor $S(Q)$ (squares) for this case. It is important to note that the first diffraction peak of $S(Q)$ occurs at a relatively smooth tail part of $\overline{P}(q)$; this implies that the interacting peak in the SANS intensity distribution is primarily reflecting the width of the first diffraction peak of $S(Q)$.

3.3. The scale intensity distribution approach

Disordered systems generally present a scattered intensity distribution consisting of a single sharp interaction peak, reflecting that a single length scale: $\Lambda = q_{\max}^{-1}$ ($q_{\max}$ is the peak position of the intensity distribution) characterize the system. It addition, the absolute intensity in a two phase-system (supramolecular aggregates like colloids and micelles) is given by a 3D Fourier transform of the Debye correlation function $\Gamma(r)$ that in this case must be of the form $\Gamma(r/\Lambda)$. Therefore,

$$I(q) = \left\langle \eta^2 \right\rangle \int_0^\infty dr 4\pi r^2 j_0(qr)\Gamma(r/\Lambda) \tag{3}$$

where $\langle \eta^2 \rangle = (1/2\pi^2) \int_0^\infty q^2 I(q) dq$ is the so-called invariant. By using $x = r/\Lambda = q_{\max} r$ and $y = q/q_{\max}$ the intensity will be:

$$q_{\max}^3 I(q)/\left\langle \eta^2 \right\rangle = \int_0^\infty dx 4\pi x j_0(xy)\Gamma(x) \tag{4}$$

a unique scaled function. On the basis of this latter equation, the plot of different scaled intensity distributions (as function of y) at different T, in the single amorphous (or liquid) structural phase, gives one single master curve. The scale intensity, on considering the intensity behavior for large q, has a simple physical meaning. Usually the onset of the Porod's law is well above the structure peak at a certain minimum wave vector q_1; therefore, by separatating the invariant in two parts, one in which is valid the Porod's law and a remaining one:

$$\frac{q_{\max}^3 I(q)}{\langle \eta^2 \rangle} = \frac{q_{\max}^3 I(q)}{(1/2\pi^2)\left[\int_0^{q_1} q^2 I(q) dq + \int_{q_1}^\infty q^2 I(q) dq\right]} \tag{5}$$

In a disordered system the Porod's contribution (second integral) to the invariant is negligible if compared with the remaining one, so that the scale intensity can be approximated as[28]:

$$\frac{q_{\max}^3 I(q)}{\langle\eta^2\rangle} \approx \frac{q_{\max}^3 I(q)}{(1/2\pi^2)\int_0^{q_1} q^2 I(q)dq} \tag{6}$$

In addition (**see e.g. inset Fig.3a**), $\overline{P(q)}$ varies slowly within the q-range of the first diffraction peak of $S(q)$, and is negligibly small beyond q_1. The scale intensity can thus be well approximated by substituting in the latter equation the general equation for $I(q)$ (**Eq.2**) and cancelling the common factor $\overline{P(q)}$ from the numerator and the denominator to obtain:

$$\frac{q_{\max}^3 I(q)}{\langle\eta^2\rangle} \approx \frac{q_{\max}^3 S(q)}{(1/2\pi^2)\int_0^{q_1} q^2 S(q)dq}, \quad for \quad q \leq q_1 \tag{7}$$

Therefore, without loosing generality, the scale intensity is proportional to the inter-particle structure factor $S(q)$ in the region of its first diffraction peak.

4. Results and discussion

4.1. Results of SANS data

Fig. 3b gives part of the theoretical fits of SANS data for the studied micellar system at different c and T. Symbols are the experimental data and lines are the fits. It is important to note that all the measured SANS spectra can be well fitted with a choice of $\epsilon = 0.03$ (with an error bar of 0.005), a value consistent with our previous study of viscosity in $L64 - D_2O$ system at high ϕ [29]. The fits in absolute intensity scale take into account the effects of the resolution and the incoherent background and give unique values of the four parameters, N, ϕ, ϵ and T^*. It can be seen that the intensity of the higher temperature liquid phase (the left-hand side peak) and the position of the peak shift toward smaller q as temperature increase. Because of the enhanced self-association (large aggregation number) as a consequence of the increased hydrophobicity of the polymer segments at higher temperatures, it can be seen clearly that the micelle grows when T increases.

The following behaviors are observed: (i) N increases increasing T and decreases as a function of c; this is due to the fact that the PPO core in the micelles becomes less hydrophilic when heated (the micelles grows as T increases for a given concentration). (ii) fitting results give also a unique function (independent from T) that relates linearly ϕ and c[17]. Furthermore, as T increases, the SANS data analysis show that the volume fraction of the micelles remains constant, for a given concentration, at all the temperatures studied. These results on ϕ are remarkable because they give the possibility to compare the experimentally determined phase diagram with that predicted by MCT. The intensity distributions in the glass region are partially resolution limited, thus we cannot perform a precise curve fitting; in addition an appropriate theory lacks for the structure factor of a system in a non-ergodic state, showing the effects of aging processes.

Fig. 4 summarizes the variations of the fitted effective temperature T^*, as a function of the temperature T for several volume fractions studied. It can be clearly seen that as T increases, the T^* increases also. There are no data shown in the glass region, which appear as gaps in the plot. **Fig. 1** shows a comparison of the phase diagram (in the $T - \phi$ plane) of the $L64 - D_2O$ micellar solution with that of the theoretical phase diagram (in the $T^* - \phi$ plane) determined by MCT for a colloidal system with a short range attraction ($\epsilon = 0.03$). In the phase diagram given in the top panel, the solid line represents the equilibrium liquid-to-crystal transition line, the dashed line the re-entrant KGT line and symbols the phase point where the SANS data are analyzed. The lower panel gives the theoretical phase diagram (MCT predicted); the phase points shown in the upper panel are then mapped into the corresponding symbols in the lower panel using results of SANS data analyses. The result of the mapping is seen to confirm the existence of the attractive branch in the predicted KGT boundary and the re-entrant liquid-to-glass-to-liquid transition.

4.2. Scaling Plots of SANS data

Fig. 5 shows a series of SANS intensity distributions and their associated scaling plots for the $c = 45wt\%$ sample, at $\phi = 0.522$, at different temperatures. In view of the fact that the $2D$ SANS patterns shows an isotropic ring, we conclude that sample stays amorphous in the entire T range studied. Looking carefully at the spectra as a function of T (top panel), there are already clues showing the KGT: judging from the width of the peak, the intensity distribution can be categorized into two groups a sharp one and a broad one. The situation become clear by analyzing the spectra using the scaling approach and taking into account that the observed $I(q)$ single peak reflects the position and the height of the structure factor first diffraction peak. Furthermore, the position and the corresponding height of the first $S(q)$ peak reflect the mean inter-particle separation of the system and the degree of local order surrounding a typical particle, respectively. Thus the height and the width of the scaling peak can be used to visualize the degree of the system local order in an amorphous state. Therefore, a sudden sharpening at a given temperature signals the onset of the liquid-to-amorphous solid transition. As one can see from the two lower panels there are two degrees of disorder which depend on T. While the narrow peak, resolution limited, represents the glassy state, the broader one (much broader than the resolution) represents the liquid state, with a broader distribution of the interparticle distance. This figure indicates that the system shows a re-entrant liquid-to-glass-to-liquid transition as temperature increases. The sharpness of the scaling peaks which are resolution limited indicates that the nearest neighbor distance in the glassy state (ranging from $306K$ to $321K$) is more uniform than that in the liquid state ($228K$ to $303K$, $325K$ to $343K$). The inset gives the peak height of the scaling plots as a function of T. The variation in the peak heights indicates the re-entrant liquid-to-attractive glass-to-liquid transition.

SANS intensity distributions and their scaling plots for a sample at $c = 48.5wt\%$ or $\phi = 0.532$, at a series of temperatures ranging from $291K$ to $340K$ are given in **Fig.6a**. The scenario is similar to the previous example: a T dependent degree of disorder characterizes the system. As T rises, the system experiences a re-entrant liquid-to-glass-to-liquid transition. However, in addition to all the similarities, when the temperature increases to $340K$, the system is driven into another glassy state peaked at $q = 0.082\text{Å}^{-1}$ From the scaling plots given in the bottom two panels, one can tell differences between these two glasses. While the narrowest peak ($340K$) is resolution limited, the slightly broader peak ($298K$ to $322K$) is also nearly resolution limited, but lower in the scaled intensities. Since difference in local structure of different states is reflected in their scaling plots, we conclude that the degree of disorder is different for these two amorphous states.

It can be interpreted that, on varying T, the system shows à liquid-to-glass-to-liquid-to-glass transition. **Fig.6b** gives the same plots taken at $c = 51.5wt\%$ or $\phi = 0.536$, for different temperatures. Generally it delivers the same message as **Fig.6a** (it shows the re-entrant transition). However, it can be see that the transition temperature between different states, for these two different ϕ samples are different. Especially in the high temperature region. For example, in the case $\phi = 0.536$, the high T liquid state is only observed at $331K$. When T increases to $333K$, the system transits to the repulsive glass state. In order to make this point clearler, the peak heights of the scaling plots as a function of T are given in the corresponding insets. The transition temperatures between different amorphous states can be visualized clearly from them. These two figures (7&8) give a firm evidence of the glass-to-liquid-to-glass transition which is in good agreement with MCT predictions for AHS systems.

According to the MCT calculations for AHS colloids (with a sufficiently short-range attraction)m, a glass-to-glass transition is predicted[30]. Although the transition between different amorphous glassy states are not uncommon in pure substances[31], yet there is no detail investigation in of the glass-to-glass transition in a colloidal-like system so far except for some recent reports on glass-to-glass transition[17,32]. From SANS intensity distributions taken at $\phi = 0.538$, given in **Fig.7**, one can see that the much broader peak (liquid state) disappears and temperature variations trigger the transition between the two amorphous solid states with different degree of disorder. By increasing T, the variation of the peak heights of the scaling plots given in the inset show a re-entrant repulsive glass-to-attractive-glass-to repulsive-glass transition. As previously said, the most important prediction of MCT for AHS system is the existence of the end point of this glass-to-glass transition line, the A_3 singularity. In the case of $\epsilon = 0.03$, the volume fraction the A_3 point is predicted to be $\phi(A_3) = 0.544$ The SANS intensity distributions and its associated scaling plots are shown in **Fig. 8a**. As one can see, all the scaling intensity curves collapse into one single master curve, indicating the fact that the first diffraction peak of $S(q)$ for all states are identical. The inset shows that all the scaling peaks have the identical height (~ 140), indicating two glasses with the same degree of local order. Increasing ϕ further to 0.546 **(Fig.8b)**, all the scaled intensities are again characterized by a unique T-dependent length scale and collapse into one single master curve showing an identical local structure of the two glasses. In other words the situation remains the same as in the previous volume fraction. This is a compelling proof that the MCT predictions are accurate.

4.3. Intermediate scattering functions (ISF)

PCS data can give further confirmations, from the dynamical point of view, on the proposed SANS results, and in particular on the novel singularity at A_3 point. In **Fig. 9** the ISF's measured at six different volume fractions ($\phi = 0.525, 0.535, 0.538, 0.542, 0, 544$ and 0.546) are shown as a function of temperature. In **Figg. 9a** and **9b** the ISF measured at $\phi = 0.525$ and 0.535 respectively, where the liquid-to-attractive glass transition is predicted, as a function of T. In the liquid state, the long time limits of the ISFs f_q^c are zero, whereas in the attractive glass state f_q^c is about 0.5, which is the DWF of the attractive glass state. The structural arrest is thus characterized by a discontinuous change in the f_q^c called a bifurcation transition. The observed occurrence of a region of a logarithmic relaxation at intermediate time, preceding the plateau region for the system in the non-ergodic state just before the transition, is highlighted by a straight line in the log-lin plot. Upon increasing ϕ to 0.542 (and 0.538), **Figg. 9c** and **9d**, respectively, we see that all the ISF can be grouped into two distinct sets of curves having two different values of DWF, one at 0.5 (attractive glass) and the other at 0.4 (repulsive glass). According to MCT, there is a possibility to observe glass-to-glass transition by varying T^*. Because $-u$ is temperature-dependent and increases on heating, T^* actually decreases as T rises, making the transition from the repulsive glass region to the attractive glass region. By comparing the long-time limit of the ISF's with the MCT predictions, we can identify the two different types of glasses by the respective DWFs. The reason for observing two different DWF's values can be interpreted as the different degree of localization of density fluctuations in the two glasses. These figures, combined with **Fig. 7** give firm evidence of the repulsive glass ($f_q^R \sim 0.4$) to-attractive glass ($f_q^A \sim 0.5$) transition. ISF's in **Fig.9e** and **Fig.9f** indicate that at A_3 point and beyond, the DWF of the two glasses states becomes identical ($f_q \sim 0.46$). These figures share the same features. It hints at critical point-like characteristics of the A_3 point. This can be considered as a definitive proof of the existence of the A_3 point singularity in the phase diagram occurring exactly at the volume fraction predicted by MCT. It is however worth noting that at the A_3 point and beyond, the intermediate time relaxations (the β relaxation region) of the two glassy states are clearly different, in spite of the fact that the long-time relaxation becomes identical. This is the first experimental finding of this interesting point that was perhaps not predicted by MCT.

5. Conclusions

We have used PCS to verify the fact that $L64D20$ micellar system have the complex phase diagram and follows the overall structural arrest transition behavior predicted by MCT for attractive colloids. In particular we show experimentally, the existence of a liquid-to-glass-to-liquid re-entrant transition, a glass-to-glass transition (associated cusp-like singularity) and A_3 singularity in the predicted phase diagram[32]. Our SANS experiments further show that, while the local structures of the attractive and the repulsive glasses are in general different, they become identical at the A_3 singularity. However, PCS results indicates that the relaxation of the two glasses are different in the intermediate temporal region. The main result of this report is the use of the SANS method, that allowing a precise fitting of the corresponding spectra in absolute unities, makes possible to pinpoint the exact volume fraction where the point of the cusp singularity and the A_3 are located[30] and to map out the whole structural arrest transition boundaries in the studied systems. From the SANS experiments, a significant difference in the local structure factor before and after the KGT is detected at various temperatures for all the volume fractions. The reason behind this experimental fact still remains unclear. It is our conjecture that the reported SANS data is reflecting the aging effects of the sample on the time scale of our measurements. We are in the process of investigating the time evolution of SANS intensity distributions at the present time. Another important suggestion comes out from reported experiments and regards the criticality of the A_3 point: it is intriguing to speculate the extent to which one can draw the analogy between this singularity and the ordinary equilibrium critical point.

6. Acknowledgments

The research at MIT is supported by a grant from Materials Science Division of US DOE. The research in Italy is supported by the PRIN2002-MIUR project.

REFERENCES

1. see e.g.: *The Physics of Complex Systems,* edited by F. Mallamace and H.E. Stanley (IOP, Amsterdam 1997); *Structure and Dynamics of Strongly Interacting Colloids and Supramolecular Aggregates in Solutions,* edited by. S. H. Chen, J. S. Huang and P. Tartaglia (Kluwer, Dordrecht 1992).
2. A.J. Liu and S. R. Nagel, Nature (London) **396**, 6706 (1998). M. E. Cates, J.P. Wittmer, J-P. Bouchaud and P. Claudin, Phys Rev. Lett., **81** 1841 (1998).
3. S. F. Edwards, Physica , **249A**, 226 (1998); S. F. Edwards and C.C. Mounfield , Physica , **226A**, 1 (1996).
4. C.H. Liu and S.R. Nagel, Phys Rev. Lett., **68** 2301 (1992); Phys Rev.B., **48** 15646 (1995).
5. M. Nicodemi, A. Coniglio. H.J. Harrmann, Phys Rev.E., **59** 6830 (1999); Physica A, 296, 451 (2001); Phys Rev. Lett., **82** 916 (1999).
6. J. Bergenholtz and M. Fuchs, Phys. Rev. E **59**, 5706 (1999);
7. L. Fabbian, W. Gœtze, F. Sciortino, P. Tartaglia, and F. Thiery, Phys. Rev. **E 59**, R1347 (1999).
8. K.A. Dawson, Curr. Op. in Coll.&Interf. Sci., **7**, 218 (2002).
9. T.G. Mason and D.A. Weitz,Phys. Rev. Lett. **75**, 2770 (1995)
10. F. Mallamace, M. Broccio, P.Tartaglia, W.R. Chen, A. Faraone and S.H. Chen, Physica A (in press).
11. F. Mallamace, P. Gambadauro, N. Micali, P. Tartaglia, C. Liao and S.H. Chen, Phys. Rev. Lett., **84** 5431 (2000).
12. T. Eckert and E. Bartsch, Phys. Rev. Lett., **89** 125701 (2002).
13. P.N. Segre, V. Prasad, A.B. Schofield and D.A. Weitz, Phys. Rev. Lett., **86** 6042 (2001); E. R. Weeks, J. C. Crocker, A. C. Levitt, A. Schofield, and D. A. Weitz, Science 287, 627 (2000).
14. A. M. Puertas, M. Fuchs and M. E. Cates, Phys. Rev. Lett., **88** 098301 (2002).
15. K.N. Pham, A.M. Puertas, J. Bergenholtz, S.U. Egelhaaf, A. Moussaid, P.N. Pusey, A.B. Scofield, M.E. Cates, M. Fuchs and W.C. Poon, Science, 296, 104 (2002).
16. M. Kapnistos, D. Vlassoppoulos, G. Fytas, K. Mortensen, G. Fleischer and J. Roovers, Phys. Rev. Lett., **85** 4072 (2000).
17. W.R. Chen, S.H. Chen and F. Mallamace, Phys. Rev. E, **66** 021403 (2002).
18. J.B. Knight, G.C. Fandrich, C.N. Lau, H.M. Jaeger and S.R. Nagel, Phys. Rev. E, **51,** 3957 (1996); Phys. Rev. E, **57**, 1971 (1998)..
19. W. Götze, in *Liquids,·Freezing and the Glass Transition*, edited by J. P. Hansen, D. Levesque, and J. Zinn-Justin (North Holland, Amsterdam, 1991).
20. L. Sjögren, J. Phys.: Condens. Matter **3**, 5023 (1991).
21. S.H. Chen, C.Y Ku and Y.C. Liu, in ref. 1 pp 243.; Eur. Phys. J. E. **9**, 283 (2002).

22. J. K. G. Dhont, J. Chem. Phys. **103**, 7072 (1995); Phys. Rev. Lett. **77**, 5304 (1995).
23. E. Bartsch, M. Antonietti, W. Shupp and H. Sillescu, J. Chem. Phys. **97**, 3950 (1992).
24. L. Lobry, N.Micali, F. Mallamace, C.Liao and S.H. Chen, Phys. Rev. E. **60**, 7076 (1999).
25. K.Z. Zhang, B. Lindmann and L. Coppola, Langmuir **11**, 538 (1995).
26. W. van Megen and S. M. Underwood, Phys. Rev. E **49**, 4206 (1994).
27. C. Liao, S.M. Choi, F. Mallamace and S.H. Chen, J. Appl. Crystallogr.**33**, 677 (2000).
28. W.R. Chen, F. Mallamace, C.J. Glinka and S.H. Chen , Phys. Rev. E, in press (2003).
29. Y.C. Liu, S.H. Chen and J.S. Huang, Phys. Rev. E. **54**, 1698 (1996).
30. K. A. Dawson, G. Foffi, M. Fuchs, W.Götze, F. Sciortino, M. Speri, P. Tartaglia, Th. Voigtmann and E. Zaccarelli, Phys. Rev. E **63**, 11401 (2001).
31. E.G. Ponyatovsky and O I Barkalov, Mater. Sci. Rep. **8**, 147 (1999).
32. S.H. Chen , W.R. Chen, and F. Mallamace, Science, **300** 619 (2003).

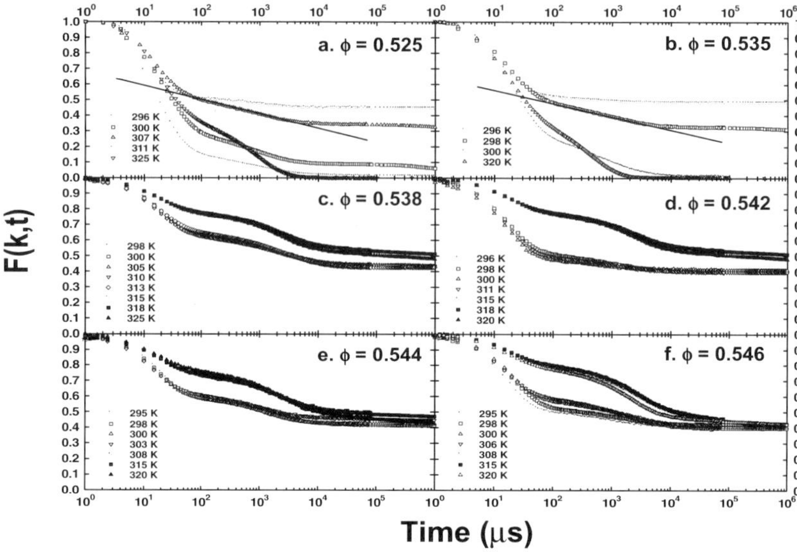

Figure 9. The ISF of different ϕ ($\phi = 0.525, 0.535, 0.538, 0.542, 0, 544$ and 0.546) are shown as a function of T in a log-lin plot. In **Fig. s 9a** and **9b** are reported the ISF's measured at $\phi = 0.525$ and 0.535 respectively, where the liquid-to-attractive glass transition is predicted. In the liquid state, the long-time limits of the ISFs f_q^c are zero, whereas in the attractive glass state f_q^c is about 0.5, which is the DWF of the attractive glass state. The logarithmic relaxation at intermediate times is highlighted by a straight line. Upon increasing ϕ to 0.542 (and 0.538), **(9c)** and **(9d)**, respectively, all the ISF's can be grouped into two distinct sets of curves having two different values of DWF, one at 0.5 (attractive glass) and the other at 0.4 (repulsive glass). According to MCT, we observe the glass-to-glass transition by varying T^*. By the comparison of these ISF's results with the MCT predictions, we identify the two different types of glasses by the respective DWFs. The reason for observing two different DWF's values can be interpreted as the different degree of localization of density fluctuations in the two glasses. These figures combined with **Fig. 7** give firm evidence of the repulsive glass ($f_q^R \sim 0.4$) to-attractive glass ($f_q^A \sim 0.5$) transition. ISF's in **Fig.9e** and **Fig.9f** indicate that at A_3 point and beyond, the DWF of the two glasses states becomes identical ($f_q \sim 0.46$).

Unifying Concepts in Granular Media and Glasses
A. Coniglio, A. Fierro, H.J. Herrmann and M. Nicodemi (editors)

Short-ranged attractive colloids: What is the gel state ?

E. Zaccarelli[a], F. Sciortino[a], S.V. Buldyrev[b], and P. Tartaglia[a]

[a]Dipartimento di Fisica and INFM Center for Statistical Mechanics and Complexity, Università di Roma "La Sapienza", Piazzale Aldo Moro 2, 00185 Roma, Italy

[b]Center for Polymer Studies and Department of Physics, Boston University, Boston, MA 02215, USA.

We evaluate thermodynamic, geometric and dynamic properties of a short-ranged square well binary mixture to provide a coherent picture of this simple, but rich, model for colloidal interactions. In particular, we compare the location, in the temperature-packing fraction plane, of the geometrical percolation locus, the metastable liquid-gas spinodal and the glass transition lines. Such comparison provides evidence that the gel-state can not be related to the attractive glass transition line directly. Indications are given for the possibility of an indirect link between the two, via an arrested phase separation process. We finally discuss the possibility that a spherical short range attraction may not be sufficient to produce an equilibrium cluster phase at low packing fraction and low temperatures.

1. Introduction

Colloidal dispersions are a suitable class of matter for many scientific purposes. Indeed, these systems are experimentally accessible with light scattering techniques and microscopy, due the the large length scales and time scales involved. Also, the inter-particle interactions can be tuned almost *ad-hoc*, for example by changing the solvent, grafting the particles or adding polymers in the dispersion. Interaction ranges much shorter than the characteristic ones of molecular or atomic liquids can be produced in colloidal suspensions. The interaction range can be reduced to a few per cent of the colloidal particle diameter. Hence, colloids can be used to test a large variety of theoretical models, or vice versa theories can be subjected to stringent experimental tests. Experimentally, the most common realization of short-range attractive colloids is obtained by the addition of small non-adsorbing polymers in the colloidal solution. These, for sufficiently small sizes, can be integrated out of the description[1], and their net effect is to produce an effective attraction on the colloidal spheres, via depletion interactions[2]. The size and the concentration of the polymers control respectively the range and the magnitude of the attraction. Theoretical models of an effective one-component attractive potential are often used, to mimic the experimental situation.

The thermodynamics of short-ranged attractive colloidal systems has been studied in great details[3,4]. By tuning the inter-particle interactions, it has been shown that, for spherical hard-sphere colloids, the addition of a particularly narrow range of the attractive part of the potential, with respect to the hard-core diameter, can produce an interesting modification of typical (*à la* Van der Walls) equilibrium phase diagrams. On decreasing the range of interaction of the attractive potential, the liquid-gas coexistence curve becomes metastable with respect to the crystal, resulting in the disappearance of the liquid as equilibrium phase.

In recent years, the study of the dynamic behaviour of short-ranged attractive colloids has revealed exciting new results. Thanks to theoretical work based on the Mode Coupling Theory (MCT) for the glass transition [5], applied to Baxter sticky spheres [6] and to the short ranged attractive Yukawa model[7], some previous astonishing experimental results[8,9] have been revisited and interpreted. MCT theoretical predictions have provided a coherent picture [10–15] of the dynamic features characteristic of short-ranged attractive potentials and have stimulated novel experiments [16–21] and simulation studies [22–28]. Near the structural arrest dynamics displays far more richness than the one observed in simply repulsive, or long-ranged attractive systems. Many predictions have been confirmed by multiple evidences, some have been questioned, others are still under investigation.

So far, most of the attention of the scientific community has addressed questions regarding the behaviour of short-ranged attractive colloidal systems in the very dense regime of colloidal particles, where the most striking predictions of MCT are manifested. In particular, it is now recognized, with the help of experiments and numerical simulations, that, at high volume fractions, two different mechanisms for glassification exist, one controlled by the excluded volume, commonly referred as 'hard-sphere' (repulsive) glass transition, and one dominated by the attractive interactions, or 'bonding', between particles, commonly termed 'attractive' glass transition. The former glass scenario is only observed at high packing fraction ϕ, while the latter is manifested for large strength of attraction, i.e. low temperatures or, in the depletion picture, large polymer concentration. These two mechanisms effectively compete with each other[29], when the range of the attractive part of the potential becomes sufficiently short with respect to the hard-core of the particles, giving rise to a reentrant fluid region between the two glasses. The system remains liquid even for packing fractions where a hard sphere repulsive system would be glassy. In MCT, this competition arises from the presence of a higher-order singularity in the control-parameter space [30,31], which regulates the anomalous dynamical behaviour in these systems, giving rise to an intriguing logarithmic decay of the density auto-correlation functions, as well as a power-law sub-diffusive behavior for the mean squared displacement. For the square well model, a recent study has provided evidence of the presence of such point[28].

However, not all predictions share the same robustness with respect to the approximations present in the MCT description. In particular, the theory is *ideal*, in the sense that it does not take into account the so-called hopping processes, that, as well known, become relevant on approaching the glass transition. These processes allow for residual diffusion where the theory would predict a complete arrest. Indeed, a recent work [27] has investigated the question of the existence of a pure glass-glass transition, as well as the stability of the attractive glass, in relation to activated bond-breaking processes.

Experimentally, dynamical arrest phenomena in short range attractive colloids are observed not only at high density, as discussed above, but also in the low packing fraction region. In this case, the arrested material is commonly named a gel. In some cases, the gel phase appears to be contiguous to a cluster phase[32–35], characterized by large supra-particular aggregates, diffusing through the sample in ergodic dynamics. The gel state displays peculiar features like the appearance of a peak in the static structure factor, for very large length scales (of the order of several particle diameters), that is stable in time, as well as a non-ergodic behaviour in the density correlation functions and a finite shear modulus. These solid-like, disordered, arrested features have induced to the appealing conjecture that these colloidal gels can be viewed as the low-density expression of the attractive glasses, both being driven by the same underlying mechanism of arrest. Support to these ideas was found in MCT itself. Indeed, the theory predicts that the attractive glass line extends, practically flat, toward very low packing fractions, almost touching at the critical point the spinodal curve, and following it on its left-hand side. The predicted large values of the non ergodicity parameter along the attractive glass line, similar to the ones observed in the gel phase, and the possibility of modeling the ergodic to non-ergodic transition locus with the attractive MCT line provide support for such an identification [7,36]. Still, some inconsistencies in the interpretation of the gel state as attractive glass have been noted. For example, in the low packing fraction regime, an identification of the attractive glass line with the gel line theory would predict a gel made of particles with average number of bonds less than two [11].

In this paper, we present Molecular Dynamics simulations of a simple model of short-ranged potential. In the high packing fraction region, the dynamics of the model displays two different arrest mechanisms, and indeed two different glass transition lines have been located. Here we complement the high density dynamic data with an evaluation of the percolation locus and of the liquid-gas spinodal. We study the intersection between the glass line and the spinodal and show that the former meets the latter at low temperatures, on the high density side. Similarly, no relation is found between the percolation locus and the glass line[37,38]. The outcome of our studies is quite unexpected, and might be crucial for understanding the nature of the colloidal gels.

2. Model, Theory and Simulation

We perform Molecular Dynamics simulations of a 50%–50% binary mixture of $N = 700$ hard spheres of mass m, with diameters σ_{AA} and σ_{BB} and ratio $\sigma_{AA}/\sigma_{BB} = 1.2$. The hard core between particles of different type $\sigma_{AB} = 0.5(\sigma_{AA} + \sigma_{BB})$. The hard core potential is complemented by an attractive square well potential of depth $-u_0$, whose well-width Δ_{ij} is controlled by varying the parameter $\epsilon = \Delta_{ij}/(\Delta_{ij} + \sigma_{ij})$, which is the same for any i, j. In this model, a 'bond' between two particles is unambiguously defined when their interaction energy is equal to $-u_0$. The simulation is based on a standard event-driven algorithm, implemented for particles interacting with SW potentials[39]. Distances are measured in units of σ_{BB}, while energy and temperature are measured in units of u_o, i.e. $k_B = 1$. Time is measured in units of $\sigma_{BB} \cdot (m/u_0)^{1/2}$, mass in units of m.

For the particular case $\epsilon = 0.03$, results of extensive simulations for this model have been reported in the high density region, in order to make comparison with the existing MCT predictions. This was carried forward in a recent attempt to build an effective mapping between theory and simulation, in order to localize and probe the dynamics near the so-called higher order singularity[28].

Ideal glass lines are predicted by MCT as ergodic to non-ergodic transitions, at which by definition the self-diffusion coefficient of atoms, or colloidal particles in our case, is equal to zero. In order to build from our simulations something comparable to an ideal glass line, we have calculated iso-diffusivity curves in the $\phi - T$ plane and studied the evolution of these curves with decreasing diffusivity. Interestingly, the shape of the ideal glass line is maintained up to quite large values of the diffusivity, allowing for a straightforward establishment of the re-entrant behaviour of the glass transition.

Combined with the iso-diffusivity curves, we present here the estimation of liquid-gas 'pre-critical' curve. Such curve is defined as the locus in the $\phi - T$ plane of points where the static structure factor $S(q) \sim 1$ at $q = 0$ and provides a close estimate to the spinodal line. Indeed, $S(0) = 1$ signals the onset of the divergence of the compressibility, i.e. it is a precursor of the phase separation into gas (colloid-poor) and liquid (colloid-rich) phases.

The square well model is also very well suited for defining a percolation threshold. Indeed, as discussed above, the existence of a bond can be defined unambiguously, when the pair interaction energy is $-u_0$. We report here the bond percolation line, beyond which space-spanning clusters of bonded particles are present in the system at a given instant. We estimated it by calculating the cluster connectivity. From an operational point of view, we have defined percolating a state point where more than 50% of the examined independent configurations (over a total of sixty) displayed an infinite cluster, spanning across the simulation box.

3. The Phase Diagram

We plot in Fig.1 the ideal MCT glass lines (GL) for the studied binary SW system with $\epsilon = 3\%$ calculated theoretically using the Percus-Yevick (PY) structure factor[28]. We also report the locus of constant diffusivity, evaluated from MD simulation, with D/D_0 [40] values varying between $5 \cdot 10^{-3}$ and $5 \cdot 10^{-6}$ together with the locus of zero diffusivity (symbols), obtained extrapolating isothermally the ϕ-dependence of the diffusion coefficient according to power-law[28]. We also show the transformation of the ideal PY MCT glass line according to the bilinear transformation[31] $\phi \to 1.897\ \phi - 0.3922$ and $T \to 0.5882\ T - 0.225$, already discussed in [28]. The parameters of the transformation are the result of fitting the locus of zero diffusivity with the transformed MCT curves. Indeed, it is typical that both MCT and the PY solution, in a certain sense, 'overestimates' the glass transition, i.e. always predicting it earlier with respect to reality (for example at higher temperatures in super-cooled liquids or at lower packing fraction for hard spheres). However, differently from hard sphere systems, where this discrepancy requires just a shift in the packing fraction, for an attractive system we have to consider a bilinear mapping [41], which considerably affects the attractive branch of the predicted glass line.

Fig.2 shows the 'spinodal' curve, the percolation curve and the locus of constant diffusivity with D/D_0 values of $5 \cdot 10^{-2}$ and $5 \cdot 10^{-3}$, significantly extending to lower packing fractions the calculations already presented in [24]). From the data shown in Fig.2 several considerations aris:

(i) the intersection between the isodiffusivity curve and the spinodal provides an estimate of the characteristic times along the spinodal. Dynamics slows down on increasing

density. The intersection between the glass line and the spinodal is located on the high density side. Hence, in this model, the attractive glass line cannot be *directly* associated to physical phenomena taking place at low packing fractions, i.e. to the gel state.
(ii) The fact that the ideal attractive GL meets the spinodal at high densities suggests an *indirect* possibility for linking the gel state to the attractive glass. Indeed, if we call T_{cg} the temperature at which the attractive GL meets the coexistence line in the high density side[42], then quenches below T_{cg} may generate upon decomposition regions where the particle concentration is within the attractive glass phase and hence which could arrest kinetically the phase separation process, leaving the imprinting of the phase separation in the frozen structure factor of the system[43–46,35]. Such hypothesis, discussed in more details in the next section, although stimulating, unfortunately cannot explain the existence of a contiguity between an "equilibrium" cluster phase and the colloidal gel[32–35].
(iii) The static percolation line is found to start from the low density side of the spinodal, and at all studied temperatures, it remains well to the left side of the largest drawn isodiffusivity curve. This means that the percolating clusters are made of particles which are moving fast, thus the lifetime of the bonds of which the clusters are made at the percolation threshold is extremely short. A study of the lifetime of the infinite cluster[47], which may provide more precise indication of the time stability of the percolating cluster as compared to diffusional times is underway. Still, the short lifetime of the bonds and the extremely large diffusional times suggest that it is not possible to establish any connection between percolation and formation of stable aggregates. On the contrary, we can rule out the possibility that, for the short ranged $\epsilon = 3\%$ model studied here, bond percolation is connected to gelation.

Recently, Miller and Frenkel [48] evaluated the critical point, the spinodal line and the percolation locus for the sticky spheres Baxter model. The Baxter model is the limiting case of the square well model when the width of the well goes to zero and u_0 goes to infinity. The relation between the sticky parameter τ [49] and the square well parameters, i.e. $u_0/k_BT = log(1 + 1/(4\tau(1/(1-\epsilon)^3 - 1)))$, based on the equality of the second virial coefficients for the two models [50], and between the Baxter packing fraction ϕ_B and the square well one, i.e. $\phi = \phi_B(1-\epsilon)^3$, allow us to compare the phase diagram of the two systems. Miller and Frenkel [48] data, upon appropriate scaling of the variables, are also reported in Fig.2. The agreement between the pre-critical curve and Baxter spinodal allows us to estimate the critical point for the attractive well to be approximately around $\phi_c \sim 0.24, T_c/u_0 \sim 0.31$ for $\epsilon = 3\%$.

4. Phase Separation and Gels

We now examine the possibility discussed in the previous section that the gel state is an arrested phase separated system, where the high density phase concentration is in the glass phase. We perform such a study by quenching from $T/u_0 = 1.0$ within the liquid-gas unstable region. We focus for simplicity on a one-component system but, since now we are interested in an aggregation process which involves clustering of many particles, we consider a much larger system than the one studied previously. We study $N = 30000$ particles, interacting via SW potential, with varying well-width, respectively equal to

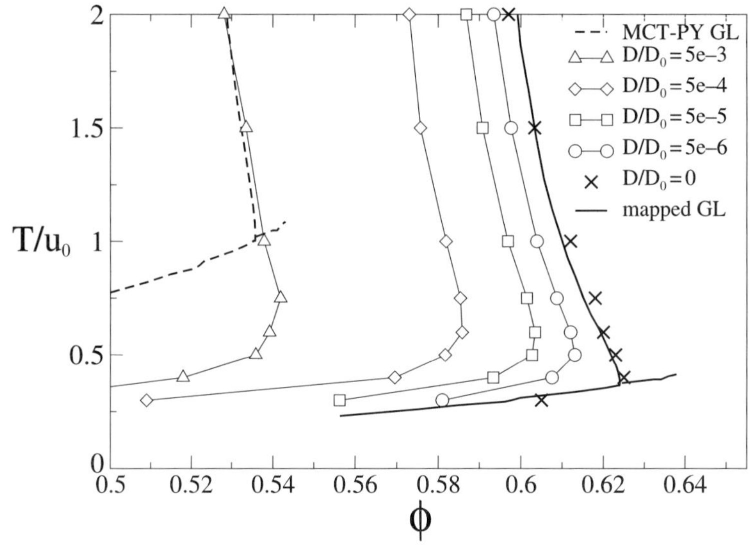

Figure 1. $(\phi, T/u_0)$-diagram in the large colloidal volume fraction region for the binary SW system with $\epsilon = 3\%$. From left to right the reported curves are: the calculated MCT glass line (GL) within Percus-Yevick approximation, iso-diffusivity curves with D/D_0 equal to $5 \cdot 10^{-3}$, $5 \cdot 10^{-4}$, $5 \cdot 10^{-5}$ and $5 \cdot 10^{-6}$, power-law extrapolated MD data from Ref.[24] for $D/D_0 = 0$ (crosses), and the MCT GL mapped as shown in Ref. [28].

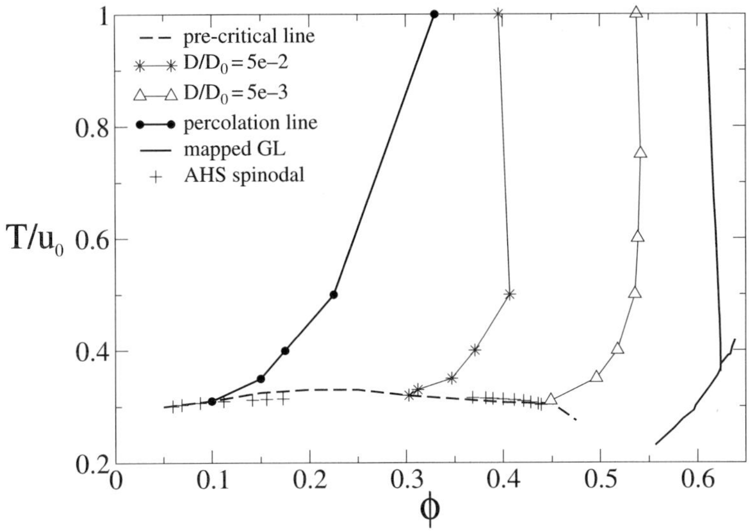

Figure 2. Same as Fig. 1, but reporting the full colloid density region. Together with the mapped GL, already shown in the previous figure, the calculated percolation and precritical lines are reported, as well iso-diffusivity curves for the values $D/D_0 = 5 \cdot 10^{-2}, 5 \cdot 10^{-3}$, both crossing the spinodal line at finite temperature. From these data, one could extrapolate the temperature where the GL meets the spinodal T_{cg} (see text). Also shown for comparison are spinodal points calculated for the Adhesive Hard Sphere (AHS) model, redrawn from [48].

$\epsilon = 0.04, 0.01, 0.005$.

We focus on the low packing fraction $\phi = 0.15$, and decrease the temperature. The choice of parameters was made in order to compare our simulation data with the phase diagram reported in [33]. Up to approximately $T/u_0 = 0.3$, the system remains in equilibrium, and the static structure factor is similar to that of a normal liquid. However, as we go further in lowering the temperature, i.e. quenching the system to $T/u_0 = 0.2$ and $T/u_0 = 0.1$, we enter in the spinodal regime and phase separation takes place. This causes particles to aggregate, producing a low wavevector peak in the static structure factor, similar, in localization and amplitude, to the one observed in [32]. However, differently from what reported in those experiments, the structure factor continues to evolve, during the simulation, although on logarithmic time-scales. This phenomenon is more marked for the largest studied well-width, i.e. $\epsilon = 0.04$, and for the highest temperature considered for the smaller widhts, $T/u_0 = 0.2$.

In Figure 3 - (upper panel), we report the time evolution of the energy per particle U/N for the various studied cases. It can be observed that there is a characteristic time-scale, which controls the aggregation kinetics. After the microscopic time-scale, there is a strong decrease in the energy, which then crosses over to a regime of very slow, approximately logarithmic, decay. The decay is slower, the narrower the studied well-width and the lower the temperature. We note that the number of bonds per particle tends to a value of about 3 or more, which, taking into account the low packing fraction of the system, indicates the formation of a cluster network. The resulting cluster phase is not an equilibrium phase, in the sense that it is driven by spinodal decomposition. To support such interpretation we show in Figure3 - (lower panel), the evolution of the static structure factor for the case $\epsilon = 0.01$ and $T/u_0 = 0.1$. The various curves represents $S(q)$ at different times from the quench, which are logarithmically spaced between $t_1 \sim 10$ and $t_2 \sim 10^4$. Times shorter than thermal equlibration are not reported. To better quantify the time dependence of the separation process, we report in Figure 4 the time evolution of the peak intensity S_{MAX} (upper panel), in analogy with the inset of Fig. 1 in Ref.[32]. For the larger well-width $\epsilon = 0.04$, the increase of the amplitude in $S(q)$, although on logarithmic time-scale, is still clearly observable after two decades in time[51]. However, for the two very narrow widths $1\%, 0.5\%$ at the extremely low temperature $T/u_0 = 0.10$ a significantly flatter behaviour is observed. Indeed, the two curves are almost superimposed onto each other at long times, and the narrower well-width shows a sharper crossover to the *quasi-plateau*.

Also, in Figure 4 - (lower panel), we plot the time evolution of the first moment of $S(q)$, i.e. $q_1 = [\sum_q qS(q)]/[\sum_q S(q)]$, which has a scaling equivalent to the peak position, but which can be calculated much more accurately [52]. These curves can be fitted at large times with a power-law $\sim t^{-\beta}$, where β varies between 0.25 and 0.5, depending on the case and on the range of fit. We recall that the typical exponent for spinodal decomposition is $1/d$, d being the dimensionality [53,45]. This result is consistent with the typical scaling of spinodal decomposition, and provides another indication that the aggregation we observe is driven by the gas-liquid phase transition. The two cases of extremely short-ranged and strong attraction display an extremely slow dynamics, which almost arrest at large times. If we return to the situation examined in the previous paragraph, we know that the attractive branch of the glass line meets the spinodal, at some very low temperature T_{cg}. It might be possible that these two cases already correspond to a temperature lower

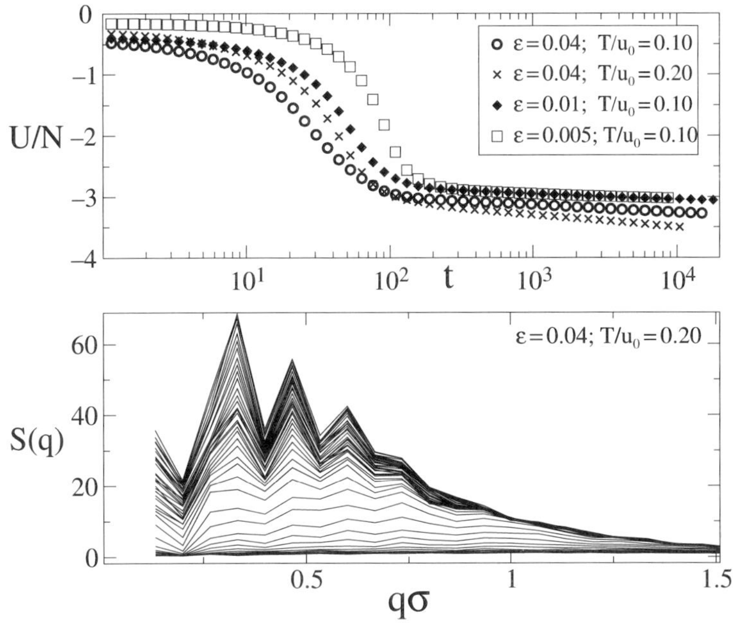

Figure 3. Upper Panel: Time evolution of the energy per particle for a one-component SW model at $\phi = 0.15$,after the quenches to $T/u_0 = 0.2$ and $T/u_0 = 0.1$. Various well-widths; Lower Panel: Static structure factors after the quench for the case $\epsilon = 0.04$ and $T/u_0 = 0.2$. The various curves refer to different times, equally spaced on a logarithmic scale. The significant noise is due to the fact that we show a single realization.

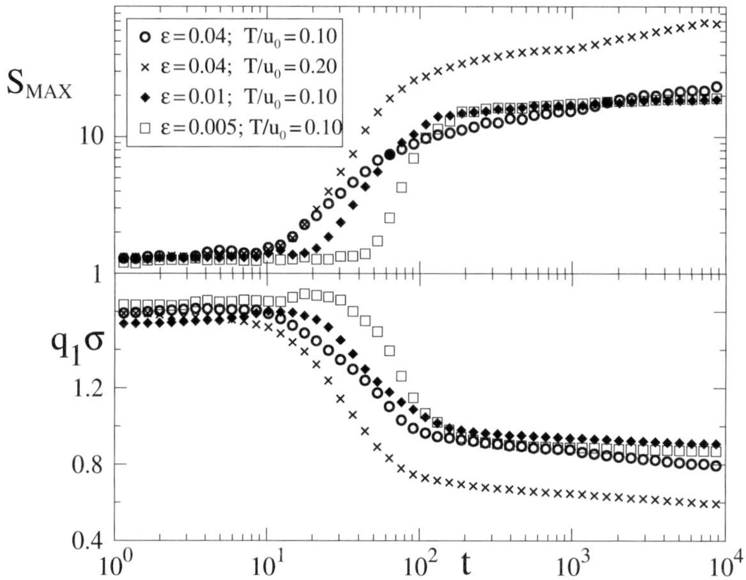

Figure 4. Upper Panel: Time evolution after the quenches of the static structure factor maximum S_{MAX}, for the same cases as in Fig. 3 (upper panel); Lower Panel: Time evolution after the quenches of the first moment q_1 in units of σ of the static structure factor.

than T_{cg}, thus what we observe here are *indirect* gels induced by spinodal decomposition.

5. Conclusions

In this paper we have provided evidence of two new important facts for simply short-ranged attractive potentials, of the kind generated via depletion interactions. Firstly, the slowing down of the dynamics in short-ranged attractive systems truly arises only at very high densities, far away from the percolation line. At low temperatures, the attractive glass line ends in the spinodal *on the high density side.* Secondly, a gel contiguous to an equilibrium cluster phase does not manifest. A gel could result from an arrested phase separation process when the density of one of the two phases crosses the attractive glass transition at low temperatures and for very short ranges of attraction ($T/u_0 = 0.1$, $\epsilon \leq 0.01$). At even lower temperatures, diffusion limited cluster aggregation[54] will take over, leaving the imprinting of the aggregation process in the gel structure[55,56].

These two observations clash with earlier conjectures that a gel state, considered as a natural extension to lower densities of the ideal MCT attractive glass, would be found in these systems [7,36]. We also note that the data reported in this article appear in disagreement with the recent theoretical model [57] (see in particular Fig. 2 of Ref. [57]), where a gel state is identified with an attractive glass after a renormalization procedure. On the other hand, our results shed some light on the fact that in real systems there must be an additional mechanism that allows to stabilize an equilibrium cluster phase. Evidence is now emerging in the scientific community that colloidal particles often tend to have a residual charge distribution [58,34,59], whose net effect is a weak, long-ranged repulsive barrier, whose importance in governing the dynamics of the system may manifest when many particles are clustered together, preventing further aggregation[60]. It would be interesting to find out if the stabilizing effect of electro-static forces implies that gels are formed by caging of clusters, instead of by caging of particles. If this were the case, particle gels would be driven by a completely different mechanism than attractive glasses, ultimately being a different form of glasses, where particles are replaced by clusters of many particles, whose size and distribution will depend on the amount of charge that is present in the colloidal suspension. Work is in progress to address specifically such possibilities.

6. Acknowledgments

We acknowledge support from MIUR PRIN, FIRB and INFM PRA-GENFDT, and also from INFM Iniziative Calcolo Parallelo. We thank W. C. K. Poon for interesting discussions.

REFERENCES

1. C. N. Likos, Phys. Rep. 348, 267 (2001).
2. S. Asakura, F. Oosawa, J. Polym. Sci. 33, 183 (1958).
3. D. Frenkel, Science 296, 106 (2002).
4. V. J. Anderson and H.N.W. Lekkerkerker, Nature 416, 811 (2002).

5. W. Götze, in Liquids, Freezing and Glass Transition, p.287, J.P. Hansen, D. Levesque and J. Zinn-Justin (eds.), North Holland, Amsterdam (1991).
6. L. Fabbian, W. Götze, F. Sciortino, P. Tartaglia and F. Thiery, Phys. Rev. E 59, R1347 (1999) and Phys. Rev. E 60, 2430 (1999).
7. J. Bergenholtz and M. Fuchs, Phys. Rev. E 59, 5706 (1999).
8. P. N. Pusey, A. D. Pirie and W. C. K. Poon, Physica A
9. E. Bartsch, M. Antonietti, W. Schupp, and H. Sillescu, J. Chem. Phys. 97, 3950 (1992).
10. K. A. Dawson, G. Foffi, M. Fuchs, W. Götze, F. Sciortino, M. Sperl, P. Tartaglia, T. Voigtmann, and E.Zaccarelli, Phys. Rev. E 63, 011401 (2001).
11. E. Zaccarelli, G. Foffi, K. A. Dawson, F. Sciortino and P. Tartaglia, Phys. Rev. E 63, 031501 (2001).
12. G. Foffi, G. D. McCullagh, A. Lawlor, E. Zaccarelli, K.A. Dawson, F. Sciortino, P. Tartaglia, D. Pini and G. Stell, Phys. Rev. E 65, 031407 (2002).
13. K. A. Dawson, G. Foffi, G. D. McCullagh, F. Sciortino, P. Tartaglia, E. Zaccarelli, Journ. Phys.: Condens. Matt. 14, 2223 (2002).
14. E. Zaccarelli, F. Sciortino, P. Tartaglia, G. Foffi, G. D. McCullagh, A. Lawlor and K. A. Dawson, Physica A 314/1-4, 539 (2002).
15. W. Götze and M. Sperl, J. Phys. Condens. Matt. 15, S869 2003.
16. F. Mallamace, P. Gambadauro, N. Micali, P. Tartaglia, C. Liao and S. H. Chen, Phys. Rev. Lett. 84, 5431 (2000).
17. K. N. Pham, A. M. Puertas, J. Bergenholtz, S. U. Egelhaaf, A. Moussaïd, P. N. Pusey, A. B. Schofield, M. E. Cates, M. Fuchs, and W. K. Poon, Science 296, 104 (2002).
18. T. Eckert and E. Bartsch, Phys. Rev. Lett. 89, 125701 (2002).
19. W.R. Chen, S.H. Chen and F. Mallamace, Phys. Rev. E, 66 021403 (2002); W.R. Chen et al, Phys. Rev. E, 68 041402 (2003).
20. S. H. Chen, W. R. Chen and F. Mallamace, Science, 300, 619 (2003).
21. K. N.Pham, S. U.Egelhaaf, P. N.Pusey, and W. C. K. Poon to be published (2003); cond-mat/0308250.
22. A. M. Puertas, M. Fuchs, M. E. Cates, Phys. Rev. Lett. 88, 098301 (2002)
23. G. Foffi , K.A. Dawson, S. Buldyrev, F. Sciortino, E. Zaccarelli and P. Tartaglia, Phys. Rev. E 65, 050802 (2002).
24. E. Zaccarelli, G. Foffi, K. A. Dawson, S. V. Buldyrev, F. Sciortino, P. Tartaglia, Phys. Rev. E 66, 041402 (2002).
25. E. Zaccarelli, G. Foffi, K. A. Dawson, S. V. Buldyrev, F. Sciortino and P. Tartaglia, J. Phys.: Condens. Matter 16,S367 (2003).
26. A. M. Puertas, M. Fuchs, M. E. Cates, Phys. Rev. E 67, 031406 (2003).
27. E. Zaccarelli, G. Foffi, F. Sciortino and P. Tartaglia, Phys. Rev. Lett. 91, 108301 (2003).
28. F. Sciortino, P. Tartaglia, and E. Zaccarelli, submitted (2003); condmat/0304192.
29. F. Sciortino, Nature Materials, News and Views, 1, 145 (2002).
30. W. Götze and M. Sperl, Phys. Rev. E **66**, 011405 (2002).
31. M. Sperl, Phys. Rev. E 68,031405 (2003); Phys. Rev. E , in press (2003).
32. P.N. Segrè, V. Prasad, A.B.chofield, and D.A. Weitz, Phys. Rev. Lett. 86, 6042 (2001).
33. V. Prasad, V.Trappe, A.D.Dinsmore, P.N.Segrè, L. Cipelletti,and D.A. Weitz, Fara-

day Discuss. 123, 1 (2003).
34. A. D. Dinsmore and D. A. Weitz, J. Phys. Condens. Matt. 14, 7581 (2002).
35. H. Segdwick, K. Kroy, M. B. Robertson, S. U. Egelhaaf and W. C. K. Poon, submitted (2003); condmat/0309616.
36. J. Bergenholtz, M. Fuchs and W. C. K. Poon, Langmuir 19, 4493 (2003).
37. M.C. Grant and W.B. Russel Phys. Rev. E 47, 2606 (1993).
38. S. K. Kumar and J. F. Douglas, Phys. Rev. Lett. 87, 188301 (2001).
39. D. C. Rapaport, The Art of Molecular Dynamic Simulation, Cambridge University Press, 1995.
40. We recall that $D_0 = \sigma\sqrt{T/m}$, is a normalizing factor, which ensures that the difference in the average velocities due to the temperature is eliminated, so that the diffusion coefficient is comparable between different temperatures.
41. This mapping is based on the hypothesis that the parameter ϵ does not require any transformation, based on the observation that it is the crucial parameter governing the anomalous behaviour of the system and the existence of the higher order singularity.
42. The ideal GL, being the outcome of the ideal MCT does not coincide with the "calorimetric" glass transition line. While in hard-spheres hopping phenomena are not significant due to the absence of an energy scale, in molecular systems it is well known that the "MCT glass" is destabilized by activated processes, which slowly restore ergodicity. Indeed, in these systems the calorimetric glass transition temperature is located well below the MCT critical temperature. In short-ranged attractive colloids, activated processes can be associated with thermal fluctuations of order u_0, which are able to break the bonds. These processes generate a finite bond lifetime and destabilize the attractive glass. Hence the effective T_{cg} must lie below the value calculated on the basis of the ideal MCT attractive GL.
43. F. Sciortino, R. Bansil, H. E. Stanley, P. Alstrom, Phys. Rev. E 47, 4615 (1993).
44. D. Sappelt, J. Jackle, Europhys. Lett. 37, 13 (1997).
45. P. Poulin, J. Bibette and D. A. Weitz, Eur. Phys. J. B 7, 277 (1999).
46. W. C. K. Poon, A. D. Pirie and P. N. Pusey, Faraday Discuss. 101, 65 (1995).
47. F. Sciortino, P. H. Poole, H. E. Stanley, and S. Havlin Phys. Rev. Lett. 64, 1686 (1990).
48. M. Miller and D Frenkel, Phys. Rev. Lett. 90, 135702 (2003).
49. R. J. Baxter, J. Chem. Phys. 49, 2770 (1968).
50. M. G. Noro and D. Frenkel, J. Chem. Phys. 113, 2941 (2000).
51. To give an estimation of this time in computer time, these simulations have required approximately 4 months in CPU time.
52. S. C. Glotzer, M. F. Gyure, F. Sciortino, A. Coniglio and H. E. Stanley, Phys. Rev. E 49, 247 (1994).
53. J. M. Gunton, M. San Miguel, P. S. Sahni, in Phase Transition and Critical Phenomena 8, 267, J.L. Lebowitz (ed.), Academic Press, London (1983).
54. For a general review see, e.g., T. Vicsek, Fractal Growth Phenomena, World Scientific, Singapore (1989).
55. M. Carpineti and M. Giglio, Phys. Rev. Lett. 68, 3327 (1992).
56. F. Sciortino and P. Tartaglia, Phys. Rev. Letts. 74, 282 (1995).
57. K. Kroy, M. E. Cates and W. C. K. Poon, condmat/0310566.

58. A. Yethiraj and A. van Blaaderen, Nature 421, 513 (2003).
59. W. C. K. Poon, private communication.
60. J. Groenewold and W. K. Kegel, J. Phys. Chem. B 105, 11702 (2001).

Unifying Concepts in Granular Media and Glasses
A. Coniglio, A. Fierro, H.J. Herrmann and M. Nicodemi (editors)

Structural arrest in chemical and colloidal gels

E. Del Gado[a], A. Fierro[b], L. de Arcangelis[c], A. Coniglio[b]

[a] Laboratoire des Verres, Université Montpellier II, France, and INFM, Napoli, Italy

[b]Dipartimento di Scienze Fisiche,Università di Napoli, and INFM, Italy

[c] Dipartimento di Ingegneria dell'Informazione, Seconda Università di Napoli, and INFM, Italy

We have introduced a minimal model for gelling systems and studied the dynamic behavior by means of numerical simulations. This study suggests a unifying picture for gelation phenomena, connecting classical gelation and recent results on colloidal systems. By varying the model parameters the slow dynamics present a crossover from the classical polymer gelation to dynamics more typical of colloidal systems, with a glassy regime that is interpreted in terms of effective clusters.

1. Introduction

At the gelation transition a viscous liquid transforms to an elastic disordered solid. In general this corresponds to the formation of a spanning structure which makes the system able to bear stress. In polymer systems the structure formation is due to chemical bonding, producing a polymerization process (chemical gelation). The gelling system typically displays critical power law behavior in the viscoelastic response and slow dynamics [1–5]: The relaxation functions show at long times a stretched exponential decay, and at the gel point the relaxation process becomes critically slow.

In colloidal systems a strong short range attraction produces a diffusion limited cluster-cluster aggregation process with a gel formation (colloidal gelation) at low density as a permanent spanning structure is formed [6–8]. The latter is generally quite different from the one of polymer gels, whereas the viscoelastic behavior is very similar. Although the attractive interaction is not able to produce a permanent gel, the formation of stable or metastable structures is detected [9–11] together with a slowing down in the dynamics. At high densities the short range attraction is able to enhance the caging effect typical of the glassy regime, and to produce a glass-like kinetic arrest at density values lower than the hard-sphere case, and depending on the strength of the attraction [12–15]. Therefore, for different values of the attraction strength and the density, colloidal systems should eventually cross over from a gel-like to a glass-like behavior. How this takes place and which is the role of the formation of stable or metastable structures is still not clear.

In order to investigate this problem we have introduced a model for gelling systems and studied the dynamic behavior by means of numerical simulations [16,17]. In the model monomers are bonded, and we consider both the cases of permanent bonds and bonds of finite lifetime (depending on the temperature) We found that with permanent bonds the system presents a structural arrest at the percolation threshold of the bonded monomer clusters. If the bond lifetime is finite, only an apparent divergence of the relaxation time is observed close to the percolation threshold, and the structural arrest occurs at higher concentrations where a glass transition is present.

Here we present the results of extensive Monte-Carlo simulations on cubic lattices. In the following sections the model and the numerical simulations are described, and the relaxation properties are studied by means of time autocorrelation functions.

2. Model and Numerical Simulations

The system we study is a solution of tetrafunctional monomers with excluded volume interactions. Each monomer occupies a lattice elementary cell and, to take into account the excluded volume interaction, two occupied cells cannot have common sites. At $t = 0$ we fix the fraction ϕ of present monomers respect to the maximum number allowed on the lattice, and randomly quench bonds between them. The four possible bonds per monomers, randomly selected, are formed with probability p_b along lattice directions between monomers that are nearest neighbors and next nearest neighbors. According to bond-fluctuation dynamics [18] the monomers diffuse on the lattice via random local movements and the bond length may vary but not be larger than l_0. The value of l_0 is determined by the self-avoiding walk condition and on the cubic lattice is $l_0 = \sqrt{10}$ in lattice spacing units.

We have first considered this model in the case of permanents bonds, i.e. once formed at $t = 0$ the bonds cannot break, nor new bonds can be formed during the dynamic evolution of the system. This corresponds to the case of chemical gelation, that can be typically obtained by irradiating the monomeric solution.

As in colloids the aggregation is due to a short range attraction and in general the monomers are not permanently bonded, we have introduced in our model a finite bond lifetime τ_b and study the effect on the dynamics. The features of this model with finite τ_b can be realized in a microscopic model: a solution of monomers interacting via an attraction of strength $-E$ and excluded volume repulsion. Due to monomers diffusion the aggregation process eventually takes place. The finite bond lifetime τ_b corresponds to an attractive interaction of strength $-E$ that does not produce permanent bonding between monomers, and $\tau_b \sim e^{E/KT}$. Due to the finite τ_b, in the simulations during the monomer diffusion the bonds between monomers are broken with a frequency $1/\tau_b$. Between monomers separated by a distance less than l_0 a bond is formed with a frequency f_b. For each value of τ_b we fix f_b so that the fraction of present bonds is always the same [16,17]. The finite bond lifetime τ_b obviously introduces a correlation in the bond formation during the simulation and may eventually lead to a phase separation between a low density and a high density phase: For the values of τ_b and f_b here considered there is no evidence of phase separation.

We let the monomers diffuse to reach the stationary state and then study the system for different values of the monomers concentration. We have considered $p_b = 1$, for which the system presents a percolation transition at $\phi_c = 0.718 \pm 0.005$. Varying the monomer concentration we have studied the autocorrelation function of density fluctuation $f_{\vec{q}}(t)$

given by

$$f_{\vec{q}}(t) = \frac{< \rho_{\vec{q}}(t+t')\rho_{-\vec{q}}(t') >}{< |\rho_{\vec{q}}(t')| >^2} \tag{1}$$

where $\rho_{\vec{q}}(t) = \sum_{i=1}^{N} e^{-i\vec{q}\cdot\vec{r}_i(t)}$, $\vec{r}_i(t)$ is the position of the $i-th$ monomer at time t, N is the number of monomers and the average $\langle ... \rangle$ is performed over the time t'. Due to the periodic boundary conditions the values of the wave vector $\vec{q}$ on the cubic lattice are $\vec{q} = \frac{2\pi}{L}(n_x, n_y, n_z)$ with $n_x, n_y, n_z = 1...L/2$ integer values. We also study the mean square displacement of the particles, $\langle \vec{r}^2(t) \rangle = \frac{1}{N}\sum_{i=1}^{N}\langle (\vec{r}_i(t+t') - \vec{r}_i(t'))^2 \rangle$. After having thermalized the system at a given value of the monomer concentration, ϕ, these quantities have been calculated, on a cubic lattice of size $L = 16$, by time averages over runs of $\sim 10^7 MCstep/particle$. The errors are calculated as the fluctuations on the statistical ensemble ($\sim 20 - 30$ different realizations of the system have been considered).

3. Relaxation properties

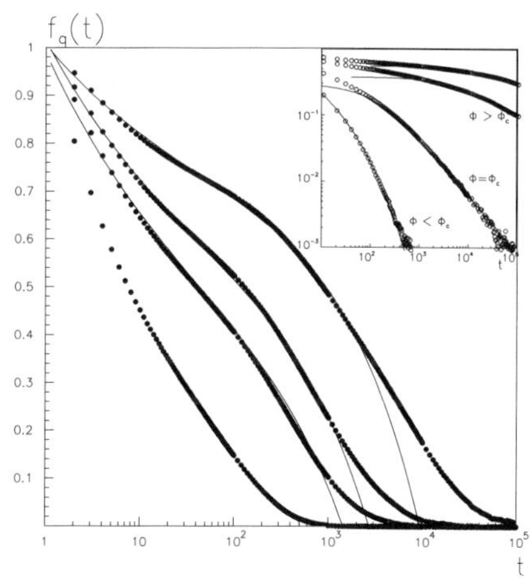

Figure 1. **Main frame** The autocorrelation functions, $f_{\vec{q}}(t)$, as function of the time for $q \sim 1.36$ for $\tau_b = 400 MCstep/particle$ (from left to right $\phi = 0.7,\ 0.8,\ 0.85,\ 0.9$). The full lines correspond to the fit with the mode-coupling β-correlator. **Inset** Double logarithmic plot of the autocorrelation functions, $f_{\vec{q}}(t)$, in the permanent bond case, for $q \sim 1.36$ and $\phi = 0.6,\ 0.718,\ 0.8,\ 0.85$. For $\phi < \phi_c$ the long time decay is well fitted by a function (full line) $\sim e^{-(t/\tau)^{\beta}}$ with $\beta \sim 0.3$. At the percolation threshold and in the gel phase in the long time decay the data are well fitted by a function $\sim (1 + \frac{t}{\tau'})^{-c}$.

In Fig.s 1 and 2, we present the time autocorrelation functions, $f_{\vec{q}}(t)$, and mean square displacement, $\langle \vec{r}^2(t) \rangle$, as function of the time for the case of permanent bonds (inset), and for the case of finite lifetime (main frame).

In the case of permanent bonds we have checked that the diffusion coefficient has reached its asymptotic limit. Starting from that time we have calculated the autocorrelation function of the density fluctuations, $f_{\vec{q}}(t)$, for ($\vec{q} = (\pi/4, \pi/4, \pi/4)$, $(\pi/2, \pi/2, \pi/2)$ and (π, π, π)), and checked that they decay to zero. Only when those two conditions are satisfied the system is considered at equilibrium. In the permanent bond case the system reaches the equilibrium over our observation time only for concentration lower than the the percolation threshold, ϕ_c. For $\phi > \phi_c$ the system remains out of equilibrium.

In the inset of Fig.1 we present the equilibrium autocorrelation functions for permanent bonds, $f_{\vec{q}}(t)$, as function of the time, for $q \sim 1.36$ ($\vec{q} = (\pi/4, \pi/4, \pi/4)$). As the monomer concentration, ϕ, approaches the percolation threshold, ϕ_c, $f_{\vec{q}}(t)$ displays a long time decay well fitted by a stretched exponential law $\sim e^{-(t/\tau)^\beta}$ with a $\beta \sim 0.30 \pm 0.05$. At the percolation threshold the onset of a power law decay is observed as it is shown by the double logarithmic plot of Fig.1 with an exponent c [2–5]. As the monomer concentration is increased above the percolation threshold in the gel phase, the long time power law decay of the relaxation functions can be fitted with a decreasing exponent c, varying from $c \sim 1.$ at ϕ_c to $c \sim 0.2$ well above ϕ_c, where a nearly logarithmic decay appears. This behavior well agrees with the one observed in gelling systems investigated in the experiments of refs. [2–5]. It is interesting to notice that this kind of decay with a stretched exponential and a power law reminds the relaxation behavior found in spin-glasses [19].

In the model with finite lifetime bonds we have checked that the average number of bonds (the average is made over different realizations of the system), has reached its asymptotic limit. Starting from that time we have calculated the autocorrelation function of the density fluctuations, $f_{\vec{q}}(t)$, for for the same values of $\vec{q}$ considered in the permanent bond case, and checked that they decay to zero. In the main frame of Fig.1 we present the equilibrium autocorrelation functions, $f_{\vec{q}}(t)$, as function of the time for $q \sim 1.36$ and $\tau_b = 400$.

At intermediate values of the monomer concentration, stretched exponential decays appear, but no onset of a power law decay is observed. For high monomer concentrations $f_{\vec{q}}(t)$ exhibits a two-step decay, that closely resembles the one observed in supercooled liquids. We fit these curves using the mode-coupling β-correlator [20], corresponding to a short time power law $\sim f + \left(\frac{t}{\tau_s}\right)^{-a}$ and a long time von Schweidler law $\sim f - \left(\frac{t}{\tau_l}\right)^{b}$, giving the exponents $a \sim 0.33{\pm}0.01$ and $b \sim 0.65{\pm}0.01$ (the full lines in Fig.1). At long times the different curves, obtained for different ϕ, collapse onto a unique master curve by opportunely rescaling the time via a factor, $\tau(\phi)$. The master curve is well fitted by a stretched exponential decay with $\beta \sim 0.50{\pm}0.06$. The characteristic time $\tau(\phi)$ diverges at a value $\phi_g \sim 0.96{\pm}0.05$ with the exponent $\gamma \sim 2.3{\pm}0.1$. This value well agrees with the mode-coupling prediction $\gamma = 1/2a + 1/2b$ [20]. A glassy pattern, characterized by a plateau, is also found in the mean square displacement, $\langle \vec{r}^2(t) \rangle$, (see Fig.2), with the diffusion coefficient going to zero as ϕ approaches ϕ_g. Analogous results are found for different τ_b, ranging from 100 to 3000$MCstep/particle$.

In Fig.3 the relaxation time, τ, calculated as $f_{\vec{q}}(\tau) \sim 0.1$, is plotted as function of the monomer concentration, ϕ, for different τ_b. For comparison we have also shown the

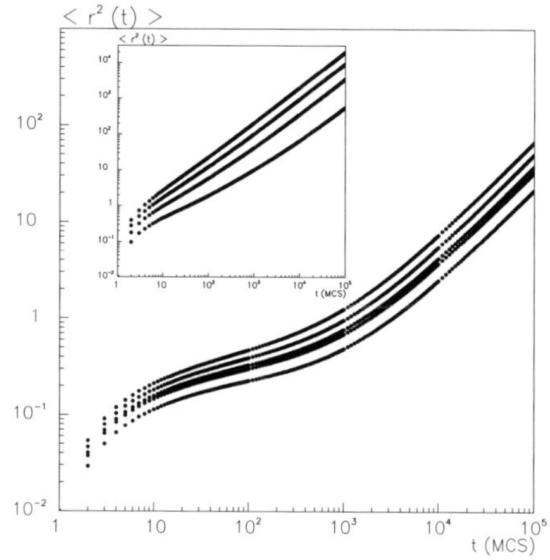

Figure 2. **Main frame** The mean square displacement of the particles, $\langle \vec{r}^2(t) \rangle$, as function of the time for $\tau_b = 1000 MCstep/particle$ (from bottom to top $\phi = 0.7,\ 0.8,\ 0.85,\ 0.9$). **Inset** The mean square displacement of the particles, $\langle \vec{r}^2(t) \rangle$, in the permanent bond case, for $\phi = 0.6,\ 0.718,\ 0.8,\ 0.85$.

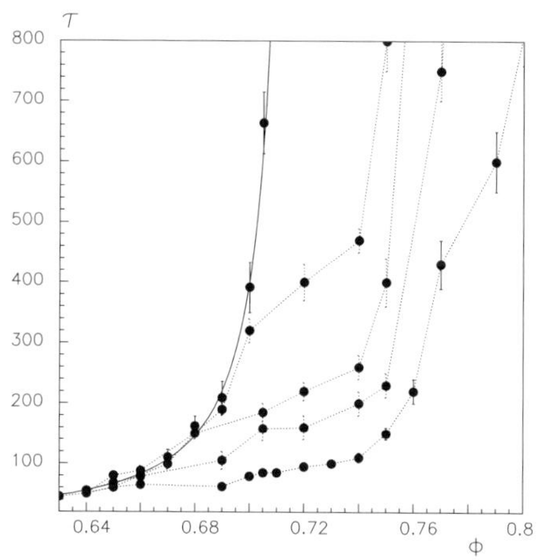

Figure 3. The average relaxation time as function of the density; from left to right: the data for the permanent bonds case diverge at the percolation threshold with a power law (the full line); the other data refer to finite $\tau_b = 3000, 1000, 400, 100 MCstep/particle$ decreasing from left to right (the dotted lines are a guide to the eye)

behavior of τ in the case of permanent bonds, which displays a power law divergence at the percolation threshold ϕ_c. We notice that for finite bond lifetime τ_b the relaxation time increases following the permanent bonds case (chemical gelation), up to some value ϕ^* and then deviates from it. The longer the bond lifetime the higher ϕ^* is. In the high monomer concentration region, well above the percolation threshold, the relaxation time in the finite bond lifetime case again displays a steep increase and a power law divergence at some higher value. This truncated critical behavior followed by a glassy-like transition has been actually detected in some colloidal systems in the viscosity behavior [21,22]. These results can be explained by considering that only clusters whose diffusion relaxation time is smaller than τ_b will behave as in the case of permanent bonds. Larger clusters will not persist and their full size will not be relevant in the dynamics: the finite bond lifetime induces an effective cluster size distribution with a cut-off, which keeps the macroscopic viscosity finite [23]. As the concentration increases the final growth of the relaxation time is due to the crowding of the particles.

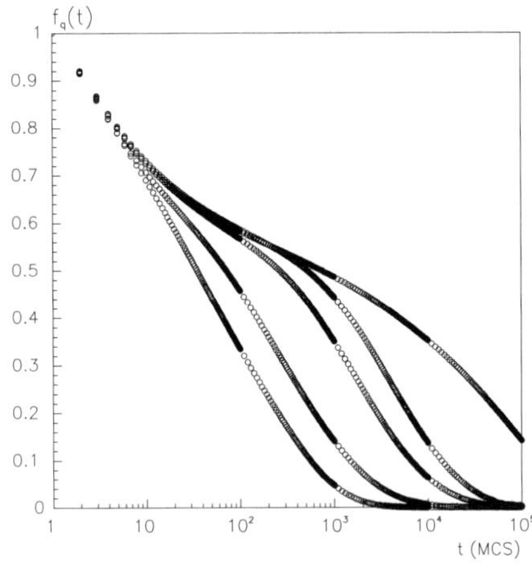

Figure 4. The autocorrelation functions, $f_{\vec{q}}(t)$, obtained for $\phi = 0.85$ and $q \sim 1.36$: the different curves refer to $\tau_b = 10, 100, 400, 1000$, compared to the permanent bond case (from left to right).

In Fig.4 we directly compare the time autocorrelation functions obtained at the same concentrations in the finite bond lifetime case and in the case with permanent bonds. We observe that for short time (of the order of τ_b) the relaxation decay coincides. This suggests that on time scales smaller than τ_b the relaxation process must be on the whole the same as in the case of permanent bonds, where permanent clusters are present in the system, and gives an interpretation in terms of the effective clusters for the two step glassy behavior of the relaxation functions: The first relaxation should be due to the motion of a cluster within the cage formed by the other clusters, whereas the second relaxation is due to the breaking of clusters. This second relaxation is the analog of the cage opening in an ordinary supercooled liquid. In conclusion, on a time scale of the order of τ_b, the effective clusters play the role of single molecules in an ordinary supercooled liquid, or in a colloidal hard sphere system. For high temperature $\tau_b \to 0$, the clusters reduce to single monomers. When τ_b is large enough (strong attraction) the cluster effect will dominate and the slow dynamics will exhibit features more closely related to chemical gelation (Fig.1). The only difference is that in the limit $\tau_b \to \infty$ we expect that the spanning cluster will have the structure of the cluster-cluster irreversible aggregation model instead of random percolation [16,17].

4. Conclusions

Our study shows that in a gelation phenomenon with permanent bond formation (chemical gelation or irreversible cluster aggregation) the divergence of the relaxation time is due to the formation of a macroscopic critical cluster and the autocorrelation function exhibits a one step decay related to the relaxation of such spanning cluster. That is, in this case the percolating cluster of bonds is the stress-bearing network which produces the divergence of the relaxation time.

In the case of non permanent bonds the relaxation time increases due the formation of effective clusters. The size of these effective clusters due to the finite lifetime does not diverge and the relaxation time exhibits a pseudo divergence corresponding to a state which we call pseudo gel or soft gel. As the density, ϕ, increases the clusters will get crowded until a glass transition is reached. In the $\tau_b - \phi$ plane therefore we have two lines: A pseudo gel line and a glass transition line. We expect that as τ_b diverges (low temperature) the two lines will end up at zero density in the cluster-cluster aggregation point. However in the low density region in general we expect that the two lines will interfere with the phase separation curve which needs to be treated with much more care. This problem is under investigation.

In conclusion, these results suggest a unifying approach for chemical gelation, colloidal gelation and colloidal glass transition. In chemical gelation and colloidal gelation the cluster formation should produce the slow dynamics. In colloidal systems for weak attraction and high concentration the system crosses over from colloidal gelation to colloidal glass due to the presence of effective clusters.

5. Acknowledgments

We would like to thank K. Dawson, A. de Candia, G. Foffi, W. Kob, F. Mallamace, N. Sator, F. Sciortino, P. Tartaglia and E. Zaccarelli for many interesting discussions. This work has been partially supported by a Marie Curie Fellowship of the European Community program FP5 under contract number HPMF-CI2002-01945, by MIUR-PRIN 2002, MIUR-FIRB 2002, CRdC-AMRA, and by the INFM Parallel Computing Initiative.

REFERENCES

1. M. Adam, D. Lairez, M. Karpasas and M. Gottlieb, Macromolecules **30**, 5920 (1997).

2. J.E. Martin, J.P. Wilcoxon and J. Odinek, Phys. Rev. A **43**, 858 (1991).
3. F. Ikkai and M. Shibayama, Phys. Rev. Lett. **82**, 4946 (1999).
4. S.Z. Ren and C.M. Sorensen, Phys.Rev.Lett. **70**, 1727 (1993).
5. P. Lang. and W. Burchard Macromolecules **24**, 815 (1991).
6. A.D. Dinsmore and D.A. Weitz J. Phys. : Condens. Matter **14**, 7581 (2002).
7. P. Meakin, Phys. Rev. Lett **51**, 1119 (1983).
8. M. Kolb, R. Botet and R. Jullien, Phys. Rev. Lett. **51**, 1123 (1983).
9. V. Trappe, V. Prasad, L. Cipelletti, P.N. Segre and D.A. Weitz, Nature **411**, 772 (2001).
10. V. Trappe and D.A. Weitz, Phys. Rev. Lett. **85**, 449 (2000).
11. P. N. Segré, V. Prasad, A. B. Schofield, and D. A. Weitz Phys. Rev. Lett. **86** 6042 (2001).
12. H. Gang, A. H. Krall, H. Z. Cummins, and D. A. Weitz, Phys. Rev. E **59**, 715 (1999).
13. F. Mallamace, P. Gambadauro, N. Micali, P. Tartaglia, C. Liao and S.H. Chen, Phys. Rev. Lett. **84**, 5431 (2000); S.H. Chen, W.R. Chen, F. Mallamace, Science **300**, 619 (2003).
14. L. Fabbian, W. Götze, F. Sciortino, P. Tartaglia and F. Thiery, Phys. Rev. E **59**, R1347 (1999).
15. E. Zaccarelli, G. Foffi, K. A. Dawson, F. Sciortino and P. Tartaglia, Phys. Rev. E **63**, 031501 (2001); A.M. Puertas, M. Fuchs and M.E. Cates, Phys. Rev. Lett. **88**, 098301 (2002).
16. E. Del Gado, A. Fierro, L. de Arcangelis and A. Coniglio, Europhys. Lett. **63**, 1 (2003).
17. E. Del Gado, A. Fierro, L. de Arcangelis and A. Coniglio, submitted to Phys. Rev. E.
18. H. P. Deutsch and K. Binder, J. Chem. Phys. **94**, 2294 (1991).
19. A. T. Ogielski Phys. Rev. B **32**, 7384 (1985).
20. W.Göetze in *Liquid, Freezing and Glass Transition*, eds. J.P. Hansen, D. Levesque and P. Zinn-Justin, Elsevier (1991).
21. F. Mallamace, S.H. Chen, Y. Liu, L. Lobry and N. Micali Physica A **266**, 123 (1999).
22. F. Laflèche, D. Durand and T. Nicolai, Macromolecules **36**, 1331 (2002).
23. A. Coniglio, J. Phys. : Condensed Matter **13**, 9039 (2001).

Unifying Concepts in Granular Media and Glasses
A. Coniglio, A. Fierro, H.J. Herrmann and M. Nicodemi (editors)

Schematic Mode Coupling Theories for Shear Thinning, Shear Thickening, and Jamming

M. E. Cates[a], C. B. Holmes[a], M. Fuchs[b], O. Henrich[b]

[a]School of Physics, The University of Edinburgh, JCMB Kings Buildings, Edinburgh EH9 3JZ, Scotland

[b]Fachbereich Physik, Universität Konstanz, D-78457 Konstanz, Germany

Mode coupling theory (MCT) appears to explain several, though not all, aspects of the glass transition in colloids (particularly when short-range attractions are present). Developments of MCT, from rational foundations in statistical mechanics, account qualitatively for nonlinear flow behaviour such as the yield stress of a hard-sphere colloidal glass. Such theories so far only predict shear thinning behaviour, whereas in real colloids both shear thinning and shear thickening can be found. The latter observation can, however, be rationalised by postulating an MCT vertex that is not only a decreasing function of strain rate (as found from first principles) but also an increasing function of stress. Within a highly simplified, schematic MCT model this can lead not only to discontinuous shear thickening but also to complete arrest of a fluid phase under the influence of an external stress ('full jamming').

1. INTRODUCTION

1.1. Arrest in Colloidal Fluids

Colloidal fluids can be studied relatively easily by light scattering [1,2]. This allows one to measure the dynamic structure factor $S(q, t_1 - t_2) = \langle \rho(\mathbf{q}, t_1)\rho(-\mathbf{q}, t_2)\rangle / N$ and also the static one, $S(q) = S(q, 0)$. Here $\rho(\mathbf{r}, t) = \sum_i \delta(\mathbf{r}_i(t) - \mathbf{r}) - N/V$; this is the real space particle density (with the mean value subtracted), and $\rho(\mathbf{q}, t)$ is its Fourier transform. For particles of radius a with short-range repulsions, $S(q)$ exhibits a peak at a value q^* with $q^*a = \mathcal{O}(1)$. The dynamic structure factor $S(q, t)$, at any q, decays monotonically from $S(q)$ as t increases. In an ergodic colloidal fluid, $S(q, t)$ decays to zero eventually: all particles can move, and the density fluctuations have a finite correlation time. In an arrested state, which is nonergodic, this is not true. Instead the limit $S(q, \infty)/S(q) = f(q)$ defines the *nonergodicity parameter*. The presence of nonzero $f(q)$ signifies frozen-in density fluctuations. Although $f(q)$ is strongly wavevector dependent, it is common to quote only $f(q^*)$ [3]. The above formulas assume time-translation invariance; nonergodic systems can violate this (showing e.g. aging phenomena) in which case $S(q, t_1 - t_2)$ as defined above must be written $S(q, t_1, t_2)$ with two time arguments.

In many colloidal materials the effective interparticle interaction $u(r)$ comprises a hard sphere repulsion, operative at separation $2a$, perhaps combined with an attraction at larger distance. (For simplicity one can imagine a square well potential of depth ϵ and range ξa, with $\xi < 1$ typically.) Colloidal fluids of this type are found to undergo nonergodicity transitions into two different broad classes of arrested nonequilibrium states. One is the colloidal glass, in which arrest is caused by the imprisonment of each particle in a cage of neighbours. This occurs even for $\epsilon = 0$ (i.e. hard spheres) at volume fractions above about $\phi \equiv 4\pi a^3 N/3V \simeq 0.58$. The nonergodicity parameter for the colloidal glass obeys $f(q^*) \simeq 0.8$. The second arrested state is called the colloidal gel. Unlike the repulsive glass, the arrest here is driven by attractive interactions, resulting in a bonded, network-type structure. Such gels can be unambiguously found, for short range attractions, whenever $\beta\epsilon \gtrsim 5 - 10$. (Here $\beta \equiv 1/k_B T$.) Hence it is not necessary that the local bonds are individually irreversible (this happens, effectively, at $\beta\epsilon \gtrsim 15 - 20$); and when they are not, the arrest is a collective, not just a local, phenomenon. It is found experimentally that for colloidal gels, $f(q^*) \gtrsim 0.9$, which is distinctly different from the colloidal glass. The arrest line for gel formation slices across the equilibrium phase diagram (e.g., plotted on the $(\phi, \beta\epsilon)$ plane), and, depending on ξ, parts of it lie within two phase regions. This, alongside any metastable gas-liquid phase boundary that is present, can lead to a lot of interesting kinetics [4,5], in which various combinations of phase separation and gelation lead to complex microstructures and time evolutions.

1.2. Mode Coupling Theory (MCT)

We do not review MCT in detail here. One widely used form of the theory [6] is based on projection methods. However, in a stripped down version (see e.g. [7,8]) the resulting equations can be viewed as a fairly standard one-loop selfconsistent approximation to a dynamical theory for the particle density field.

We take $\beta = 1$, bare particle diffusivities $D_0 = 1$, and start from the overdamped Langevin equations $\dot{\mathbf{r}}_i = \mathbf{F}_i + \mathbf{f}_i$ for independent particles of unit diffusivity subjected to external forces $\mathbf{F}_i$ and random forces $\mathbf{f}_i$. One proceeds by a standard route to a Smoluchowski equation $\dot{\Psi} = \Omega\Psi$ for the N-particle distribution function Ψ, with evolution operator $\Omega = \sum_i \nabla_i.(\nabla_i - \mathbf{F}_i)$. Now take the forces $\mathbf{F}_i$ to originate (via $\mathbf{F}_i = -\nabla_i H$) from an interaction Hamiltonian

$$H = -\frac{1}{2}\int d^3\mathbf{r} d^3\mathbf{r}' \rho(\mathbf{r})\rho(\mathbf{r}')c(|\mathbf{r} - \mathbf{r}'|) \tag{1}$$

where $Nc(q) = V[1 - S(q)^{-1}]$. This is a harmonic expansion in density fluctuations; $c(q)$ is the direct correlation function, and this form ensures that $S(q)$ is recovered in equilibrium. We neglect solvent mediated dynamic forces (hydrodynamic couplings). Also, in principle these couplings mean that the noise in the Langevin equation should be correlated between particles, in contrast to the independent white noise assumed here. In addition we neglect anharmonic terms in H; to regain the correct higher order density correlators (beyond the two point correlator $S(q)$) in equilibrium, these terms would have to be put back.

From the Smoluchowski equation (or the corresponding nonlinear Langevin equation for the density $\rho(\mathbf{r})$[7,9]), one can derive a hierarchy of equations of motion for correlators such as $S(q,t)$, more conveniently expressed via $\Phi(q,t) \equiv S(q,t)/S(q)$. Factoring arbitrarily the four-point correlators that arise in this hierarchy into products of two Φ's, one obtains a closed equation of motion for the two point correlator

$$\dot{\Phi}(q,t) + \Gamma(q)\left[\Phi(q,t) + \int_0^t m(q,t-t')\dot{\Phi}(q,t')dt'\right] = 0 \quad (2)$$

where $\Gamma(q) = q^2/S(q)$ is an initial decay rate, and the memory function obeys

$$m(\mathbf{q},t) = \sum_{\mathbf{k}} V_{\mathbf{q},\mathbf{k}}\Phi(\mathbf{k},t)\Phi(\mathbf{k}-\mathbf{q},t) \quad (3)$$

with the vertex

$$V_{\mathbf{q},\mathbf{k}} = \frac{N}{2V^2q^4}S(q)S(k)S(|\mathbf{k}-\mathbf{q}|)[\mathbf{q}.\mathbf{k}c(k) + \mathbf{q}.(\mathbf{k}-\mathbf{q})c(|\mathbf{k}-\mathbf{q}|)]^2 \quad (4)$$

Equations 2-4, which are slightly simpler than the ones used in molecular glasses because of the justified neglect of inertial terms in an overdamped environment, completely define the MCT as usually applied in colloidal systems [6].

The MCT equations exhibit a bifurcation that corresponds to a sudden arrest transition, upon smooth variation of either the density ϕ or any interaction parameters that control $c(q)$ (equivalently, $S(q)$). Here the nonergodicity parameters $f(q)$, suddenly jump (for all q at once) from zero to nonzero values. Near this (on the ergodic side, which is always the direction MCT approaches from), $\Phi(q,t)$ develops interesting behaviour. Viewed as a function of time, it decays onto a plateau of height $f(q)$, stays there for a long time, and then finally decays again at very late times. The two decays are called β and α respectively. Upon crossing the bifurcation, their relaxation times diverge smoothly with the parameters; upon crossing the locus of this divergence, $f(q) \equiv S(q,\infty)$ jumps discontinuously from zero to a finite value.

2. MCT & DYNAMIC HETEROGENEITY

It is interesting to compare the MCT approach with the concept of dynamical heterogeneity and/or 'assisted dynamics' (e.g. [10]). MCT ignores locally activated processes but treats collective density fluctuations in a relatively sophisticated way; for most theories of dynamic heterogeneity, exactly the reverse applies. Therefore, neither theory can claim to offer a complete picture. Much evidence on colloids suggests that there are indeed localized regions of enhanced mobility (e.g. [11,12]); but this very idea requires some sort of immobilized background state within which such excitations arise. MCT addresses the onset of this collectively arrested state. Ignoring the excitations may be only a modest error if the density of excitations near the MCT transition is small enough, but could undermine the whole approach if it is large. Conversely, any theory of localized defect dynamics within a frozen matrix (modelled, e.g., on a lattice [10]) should work only if, on coming from the glass side, the proliferation of defects is not pre-empted by a collective unfreezing of the matrix through an MCT-like mechanism.

The adequacy of either theory may depend on what type of glass is under study. For colloids, MCT seems surprisingly adequate [13]. A striking recent success concerns systems with both attractive interactions and hard-core repulsions. First, MCT unambiguously predicts [14–16] that adding a weak, short range attraction to the hard sphere system should melt the glass (which has $f(q^*) \sim 0.8$). Second, MCT predicts that adding more of the same attraction should mediate a second arrest, this time into a 'gel' state of high nonergodicity parameter ($f(q^*) \sim 0.95$). Third, MCT predicts that as parameters are varied, a higher order bifurcation point should enter the picture, resulting in a characteristic logarithmic decay for $\Phi(q,t)$. Although not every detail of this scenario is yet confirmed, there is clear experimental evidence for both the re-entrance, and the logarithmic decay [17,18]. The latter is also seen clearly in recent simulations [19,20]. These successes of MCT pose a notable challenge to dynamical heterogeneity theories of the glass transition in colloids. Until such a theory can explain the three features just outlined, it is fair to conclude that MCT remains the least inadequate theory of the colloidal glass transition [9].

3. SHEAR THINNING

3.1. A Microscopic Approach

In Ref.[21], a theory is propounded, along MCT lines, of colloidal suspensions under flow. The work was intended mainly to address the case of repulsion-driven glasses, and to study the effect of imposed shear flow either on a glass, or on a fluid phase very near the glass transition. In either case, simplifications might be expected because the bare diffusion time $\tau_0 = a^2/D_0$ is small compared to the 'renormalized' one $\tau = a^2/D$, which in fact diverges (enslaved to the α relaxation time) as the glass transition is approached. If the imposed steady shear rate is $\dot\gamma$, then for $\dot\gamma\tau_0 \ll 1 \leq \dot\gamma\tau$, one can hope that the details of the local dynamics are inessential and that universal features related to glass formation should dominate. Note, however, that by continuing to use a quadratic H (Eq.1), we will assume that, even under shear, the system remains 'close to equilibrium' in the sense that the density fluctuations that build up remain small enough for a harmonic approximation to be useful. This may well be inadequate for hard spheres, but a systematic means of improvement upon it is not yet available.

The basic route followed in Ref.[21] is quite similar to that laid out above for standard MCT, modulo the fact that an imposed shear flow is now present. A key simplification is to neglect velocity fluctuations so that the imposed shear flow is locally identical to the macroscopic one; this cannot be completely correct, but allows progress to be made. For related earlier work see Refs.[22,23].

We again take $\beta = 1$, $D_0 = 1$, and start from the Langevin equations $\dot{\mathbf{r}}_i = \mathbf{u} + \mathbf{F}_i + \mathbf{f}_i$ for independent particles of unit diffusivity subjected to external forces $\mathbf{F}_i$ and, now, an imposed flow velocity $\mathbf{u}(\mathbf{r}_i)$. We take this to be a simple shear flow with $\mathbf{u}(\mathbf{r}) = \dot\gamma y \hat{\mathbf{x}}$. The Smoluchowski equation $\dot\Psi = \Omega\Psi$ is unchanged but the evolution operator is now $\Omega = \sum_i \nabla_i . (\nabla_i - \mathbf{F}_i - \mathbf{u}(\mathbf{r}_i))$. So far, the adaption to deal with shearing is fairly trivial. The next stages are not. We assume an initial equilibrium state with $\Psi(t=0) \propto \exp[-H]$, and switch on shearing at $t = 0+$. We define an *advected correlator*

$$\Phi(\mathbf{q},t) = \langle \rho(\mathbf{q},0)\rho(-\mathbf{q}(t),t)\rangle / S(q)N \tag{5}$$

where $\mathbf{q}(t) = (q_x, q_y + q_x\dot\gamma t, q_z)$. This definition subtracts out the trivial part of the advection, which merely transports density fluctuations from place to place. The nontrivial part comes from the effect of this transport on their time evolution; the main effect (see e.g. [23]) is to kill off fluctuations by moving their wavenumbers away from q^* where restoring forces are weakest (hence the peak there in $S(q)$). Hence the fluctuations feel a stronger restoring force coming from H, and decay away more strongly. This feeds back, through the nonlinear term, onto the other fluctuations, including ones transverse to the flow and its gradient (i.e., with $\mathbf{q}$ along z) for which the trivial advection is absent.

There follow a series of MCT-like manipulations which differ from those of the standard approach because they explicitly deal with the switching on of the flow at $t = 0+$. We integrate through the transient response to obtain the steady state correlators, under shear, as $t \to \infty$. (There is no integration through transients in standard MCT; one works directly with steady-state quantities.) Despite all this, the structure of the resulting equations is remarkably similar to Eqs. 2,3:

$$\dot{\Phi}(\mathbf{q},t) + \Gamma(\mathbf{q},t)\left[\Phi(\mathbf{q},t) + \int_0^t m(\mathbf{q},t,t')\dot{\Phi}(\mathbf{q},t')dt'\right] = 0 \tag{6}$$

This equation describes transient density fluctuations; that is, time zero in Eq.5 corresponds to the switch-on time of a shear flow, so that Φ is a particular instance of a two-time correlator rather than a one-time one. For a related approach based on the one-time correlator of fluctuations around steady state (but assuming a fluctuation-dissipation theorem which need not apply here), see [24]. Eq.6 involves a time dependent, anisotropic "initial decay rate":

$$\Gamma(\mathbf{q},t)S(q) = q^2 + q_xq_y\dot{\gamma}t + (q_xq_y\dot{\gamma}t + q_x^2\dot{\gamma}^2t^2)S(q) - q_xq_y\dot{\gamma}S'(q)/q \tag{7}$$

The memory kernel is no longer a function of the time interval $t - t'$ but depends on both arguments separately

$$m(\mathbf{q},t,t') = \sum_{\mathbf{k}} V(\mathbf{q},\mathbf{k},t,t')\Phi(\mathbf{k},t)\Phi(\mathbf{k}-\mathbf{q},t) \tag{8}$$

through a time-dependent vertex V whose detailed derivation will appear in [25]:

$$\begin{aligned} V(\mathbf{q},\mathbf{k},t,t') = {} & [NS_kS_p/\left(2V^2S_q\Gamma(\mathbf{q},t)\Gamma(\mathbf{q},t')\right)] \\ & [\,\mathbf{q}(t).\mathbf{k}(t-t')c(k(t-t')) + \mathbf{q}(t).\mathbf{p}(t-t')c(p(t-t'))+ \\ & S(q(t-t'))\mathbf{q}(t).\left(\mathbf{q}(t-t')c(q(t-t'))-\mathbf{q}c(q)\right)] \\ & [\,\mathbf{q}.\mathbf{k}c(k) + \mathbf{q}.\mathbf{p}c(p)+ \\ & q_xq_y\dot{\gamma}t\,S(q)\left(c(k) + c(p) - \tfrac{N}{2V}(c(q)c(k) + c(q)c(p) + c(k)c(p))\right)] \end{aligned} \tag{9}$$

Here the abbreviation $\mathbf{p} = \mathbf{q} - \mathbf{k}$ is used, and advected wavevectors (see below Eq. 5) carry a time dependence.

Using a nonequilibrium Kubo-type relationship [21] one can also obtain an expression for the steady state viscosity $\eta = \sigma(\dot{\gamma})/\dot{\gamma}$ where $\sigma(\dot{\gamma})$ is the shear stress as a function of shear rate. The viscosity is expressed as an integral of the form

$$\eta = \int_0^\infty dt \sum_{\mathbf{k}} V_\eta(\mathbf{k},t)\Phi^2(\mathbf{k},t) \tag{10}$$

where the function V_η may be found in Ref. [21].

The above calculations give several interesting results. First, any nonzero shear rate, however small, restores ergodicity for all wavevectors (including ones which are transverse to the flow and do not undergo direct advection). This is important, since it is the absence of ergodicity that normally prevents MCT-like theories being used inside the glass phase, at $T < T_g$ or $\phi > \phi_g$. Here we may use the theory in that region, so long as the shear rate is finite.

In the liquid phase ($\phi < \phi_g$) the resulting flow curve $\sigma(\dot\gamma)$ shows shear thinning at $\dot\gamma\tau \gtrsim 1$, which is when the shearing becomes significant on the timescale of the slow relaxations. This is basically as expected. Less obviously, throughout the glass, one finds that the limit $\sigma(\dot\gamma \to 0+) \equiv \sigma_Y$ is nonzero. This quantity is called the yield stress and represents the minimum stress that needs to be applied before the system will respond with a steady-state flow. (For lower stresses, various forms of creep are possible, but the flow rate vanishes in steady state.)

The prediction of a yield stress in colloidal glasses is significant, because glasses, operationally speaking, are normally defined by the divergence of the viscosity. However, it is quite possible for the viscosity to diverge without there being a yield stress, for example in 'power law fluids' where $\sigma(\dot\gamma) \sim \dot\gamma^p$ with $0 < p < 1$ [27]. This does not happen in the present calculation, where the yield stress jumps discontinuously from zero to a nonzero value, σ_Y^c, at ϕ_g. The existence of a yield stress seems to be in line with most experimental data on the flow of colloidal glasses, although one must warn that experimentalists definitions of what a yield stress is, do vary across the literature [28]. Ours is defined as the limiting stress achieved in a sequence of experiments at ever decreasing $\dot\gamma$, ensuring that *a steady state is reached* for each shear rate before moving onto the next one. The latter requirement may not be practically achievable since the equilibration time could diverge smoothly at small $\dot\gamma$: certainly one would expect to have to wait at least for times t such that $\dot\gamma t \gtrsim 1$. But unless the flow curve has unexpected structure (absent in this approach) at small shear rates, the required extrapolation can presumably be made.

3.2. Schematic MCT models

It has long been known that the key mathematical structure behind Eqs. 2-4 can be captured by low-dimensional schematic models in which the full **q** dependence is suppressed [29,6]. In other words, one chooses a single mode, with a representative wavevector around the peak of the static structure factor, and writes mode coupling equations for this mode treated by itself. At a phenomenological level, one can capture the physics similarly even with shearing present (despite the more complicated vectorial structure that in reality this implies). Specifically one can define [21] the $F_{12}^{\dot\gamma}$ model — the sheared extension of a well known static model, F_{12} — via

$$\dot\Phi(t) + \Gamma\left[\Phi(t) + \int_0^t m(t-t')\dot\Phi(t')dt'\right] = 0 \tag{11}$$

with memory function (schematically incorporating shear)

$$m(t) = [v_1\Phi(t) + v_2\Phi^2(t)]/(1+\dot\gamma^2 t^2) \tag{12}$$

The vertex parameters $v_{1,2}$ are smooth functions of the volume fraction ϕ (and any interactions). To calculate flow curves, etc., one also needs a schematic form of Eq.10; here we take the first moment of the correlator to fix the time scale for stress relaxation (which is, in suitable units, simply the viscosity):

$$\eta = \int_0^\infty \Phi(t)dt \tag{13}$$

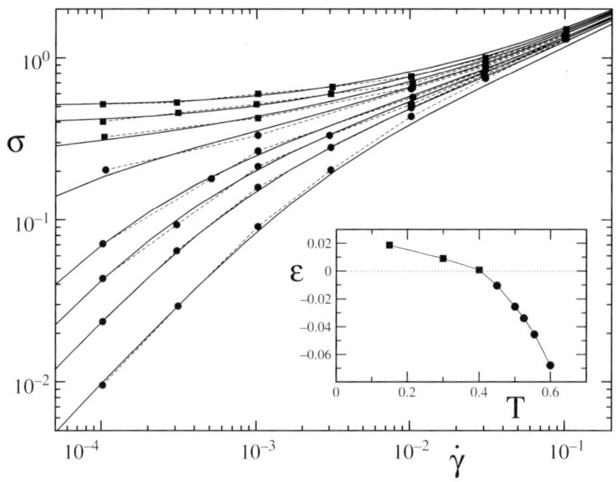

Figure 1. Symbols are shear stress data of a super-cooled Lennard–Jones binary mixture in reduced units taken from Ref. [31]; from top to bottom, the temperatures are 0.15, 0.3, 0.4, 0.45, 0.5, 0.525, 0.555 & 0.6 while the transition lies at $T \approx 0.435$. The solid lines give fits by eye using the $\mathrm{F}_{12}^{\dot{\gamma}}$–model with parameter ε measuring the separation from the transition shown in the inset. Units are converted by $\sigma = 1.5\sigma^{\text{theo.}}$ and $\dot{\gamma} = 1.3\dot{\gamma}^{\text{theo.}}$. More details and a preliminary fit can be found in [30].

(Note that a different choice, e.g. with $\Phi(t)^2$ in this equation to closer resemble Eq.10, would yield quite similar results.) This simplest of schematic models gives very similar results to a much more sophisticated (but still schematic) approximation of the full equations [21], with $\sigma - \sigma_Y \sim \dot{\gamma}^{0.16}$ and $\sigma_Y - \sigma_Y^c \sim (\phi - \phi_g)^{1/2}$. Such predictions can be compared with experiment [30] and, as shown in Figure 1, suggest that the more advanced schematic models are at least semi-quantitative.

4. SHEAR THICKENING AND JAMMING

The calculations described above predict, generically, shear thinning behaviour: advection kills fluctuations, reducing the α relaxation time, which causes the system to flow more easily at higher stresses. However, in some colloidal systems, the reverse occurs. This is shear thickening, and gives a flow curve $\sigma(\dot{\gamma})$ with upward curvature. In extreme cases, an essentially vertical portion of the curve is reported [32,33]. One interpretation of the latter scenario (called 'discontinuous shear thickening') is that the underlying flow curve is actually S-shaped. Since any part of the curve with negative slope is mechanically unstable (a small increase in the local shear rate would cause an acceleration with positive feedback), this allows a hysteresis cycle in which, at least according to the simplest models, discontinuous vertical jumps on the curve bypass the unstable section (see Figure 2).

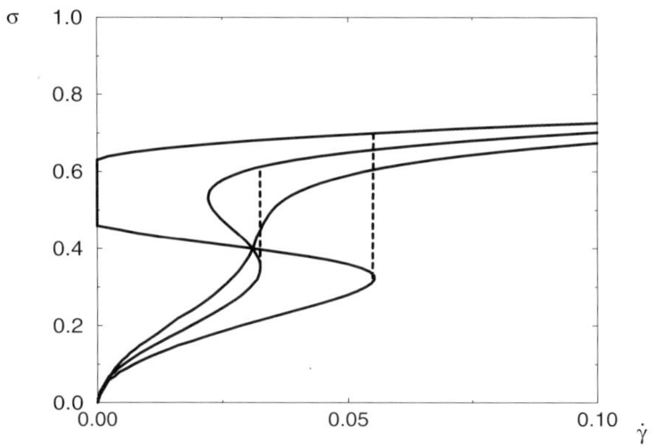

Figure 2. Three possible flow curves for a shear thickening material. The monotonic curve corresponds to continuous shear thickening. The remaining two curves are S-shaped; one expects, on increasing the shear rate, the stress to jump from the lower to upper branch at (or before) the vertical dashed line shown in each case. One curve shows the full jamming scenario: the existence of an interval of stress, here between 0.45 and 0.63, within which the flow rate is zero, even in a system ergodic at rest. (Stress and strain rate units are arbitrary.)

If this viewpoint is adopted, there seems to be nothing to prevent the upper, re-entrant part of the curve from extending right back to the vertical axis (see Figure 2) in which case there is zero steady-state flow within a certain interval of stress. The system has both an upper and a lower yield stress delimiting this region. (If it is nonergodic at rest, it could also have a regular yield stress on the lower part of the curve near the origin – we ignore this here.) This case has been called 'full jamming' [34]. Although mostly a theoretical speculation, at least one experimental report of this kind of behaviour has appeared in the literature recently [35].

The above discussion suggests that shear thickening and full jamming might be viewed as a stress-induced glass transition of some sort [36]. If so, it is natural to ask whether this idea can be accommodated within an MCT-like approach. Since the analysis of Ref. [21] gives only shear thinning, this is far from obvious. In particular, a stress-induced glass transition would require the vertex V to 'see' the stress; this might require one to go beyond harmonic order in the density, that is, it might require improvement to Eq.1. Indeed, since it is thought that jamming arises by the growth of chainlike arrangements of strong local compressive contacts [36], it is very reasonable to assume that correlators beyond second order in density should enter.

In [37] we develop a schematic model along the lines of Eqs.11–13 to address shear thickening (with, for simplicity, $v_2 = 0$). This is the $F_1^{\dot\gamma,\sigma}$ model

$$\dot\Phi(t) + \Gamma\left[\Phi(t) + \int_0^t m(t-t')\dot\Phi(t')dt'\right] = 0 \tag{14}$$

with memory function

$$m(t) = [v_0 + \alpha\sigma]\exp[-\dot\gamma t]\Phi(t) \tag{15}$$

and viscosity $\eta = \sigma/\dot\gamma$ obeying

$$\eta = \int_0^\infty \Phi(t)dt\,. \tag{16}$$

The memory function now schematically incorporates both the loss of memory by shearing and a stress-induced shift of the glass transition. (Without stress or shear, the latter occurs at $v_0 = 4$.) The choice of an exponential strain rate dependence is purely for algebraic convenience, whereas the form in Eq. 12 is closer to the one found in the full **q**-dependent vertex under shear (see section 3.1 above and [21]). The choice of a linear dependence of the vertex on stress (rather than the quadratic one that would arise in a Taylor expansion about the quiescent state) can be viewed as a linearization about a finite stress chosen to lie close to the full jamming region: this, rather than the behaviour at very small stresses, is the interesting region of the model. In any case, the qualitative scenarios that emerge from Eqs.14–16 are relatively robust to the precise details of the model [37].

This model results in a 'full jamming' scenario as part of a wider range of rheological behaviour. Fig. 3 shows three kinds of thickening behaviour, dependent on model parameters; v_0 is varied close to the quiescent glass transition, and for the chosen α there is a progression from a monotonic, continuously shear-thickening curve, via a nonmonotonic S-shaped curve, to a curve that extends right back to the vertical axis. For the largest values of the parameter v_o, in Fig. 3, there is therefore a range of stress for which the shear rate returns to zero: there is then no ergodic solution, and the jammed state is stable. This represents full jamming. Note that if, as seems likely, α depends on the details of interparticle interactions, then the evolution between these scenarios does too. This makes sense since one would certainly expect hard particles to be more 'jammable' than soft ones.

Fig.3 is qualitatively similar to Fig.2 – whose data actually comes from [34]. In that work a somewhat similar theory is developed, based not on MCT but on the trap model of glasses. The emergence of the same qualitative scenario from two quite different approaches to glass rheology is reassuring, although in each case the ansatz of a stress-dependent glass transition was, effectively, put in by hand.

The lower and upper endpoints σ_{c1} and σ_{c2} of the stable jammed state represent distinct jamming transitions. Their critical stresses obey

$$f_c\left[(v_o + \alpha\sigma_c)\,f_c - 2\right] = \sigma_c, \tag{17}$$

where f_c is given by the largest solution of $\frac{f_c}{1-f_c} = (v_o + \alpha\sigma_c)f_c^2$. Such transitions exist provided that both v_o and α are sufficiently large. Bertrand *et al* [35] found that, for concentrations below a certain value, their samples showed ordinary thickening, whilst above this value the shear-induced solid was seen. The behaviour illustrated in Fig. 3 is

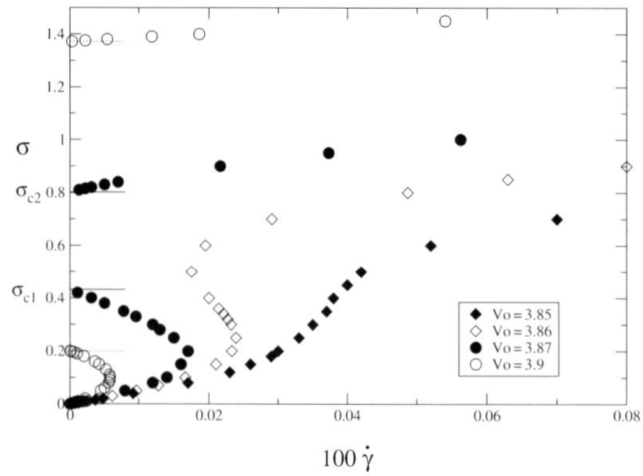

Figure 3. Flow curves for $\alpha = 0.95$. For the two largest values of v_o, it appears that for a window in σ, the relaxation time has diverged. Analytic calculations of the limits of this window are indicated as horizontal lines near the stress axis. These values of the stress are dubbed σ_{c1} and σ_{c2}, as shown here for one of the parameter sets.

reminiscent of this. Note that the re-fluidisation under increasing stress depends on α: if this is too large (for a given v_o) this re-fluidisation is not present. The resulting 'phase diagram' of the model, dividing parameter space into ergodic and nonergodic regions, is shown in Fig. 4. At large enough stresses, jammed states arise for $\alpha > 1$. However, for particle densities close to but below the quiescent glass transition, for $\alpha < 1$ the system jams in an intermediate window of stress.

5. GLASSY VERSUS HYDRODYNAMIC THICKENING

Shear thickening is widely reported (e.g. [32,33,38,39]) and usually attributed to a buildup of hydrodynamic forces between clusters of particles [40,41]. Our work suggests that, at least in some systems, this may not be the only mechanism at work. In particular, Fig. 3 admits shear thickening at Peclet numbers $\dot{\gamma}\tau_0 \sim 10^{-4}$, rather than values of order unity predicted by most theories of hydrodynamic clustering. Such theories do not so far appear to offer any natural explanation of the S shaped flow curve that appears to underly *discontinuous* shear thickening (see e.g. [42], and references therein). On the other hand, simulations of dense colloids do predict, for hard spheres in the absence of Brownian motion, a catastrophic jamming transition. In this transition, a network of close contacts propagates to infinity at *finite strain*, creating a solid [43]. The relation between this and our own model in which Brownian motion of course plays an essential role, is yet to become clear.

It should also be pointed out that, to whatever extent full jamming is actually observed

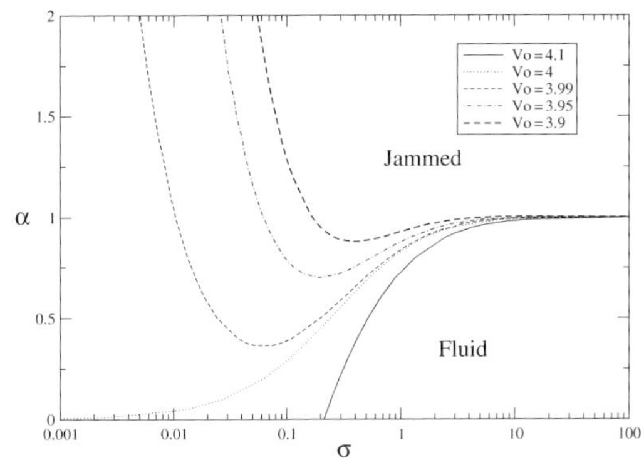

Figure 4. 'Phase diagrams' for the model for various v_o. The lines denote transitions in the (α, σ)-plane. All states below the curve for a given value of v_o are fluid states, whilst those above (and on) the line are nonergodic, jammed states.

[35], hydrodynamic theories cannot explain it. This is because hydrodynamic forces are dynamical in origin and therefore cannot be responsible for maintaining a purely static state of arrest. It is however conceivable that a limit exists in which interparticle velocities and separations both vanish at late times in such a fashion that the resulting forces approach constant values. However, we do not find this particularly plausible.

Note in any case that existing hydrodynamic theories (rather than simulations [43]) of colloid rheology, by taking no account of the glass transition, predict that the zero shear viscosity diverges only at random close packing (volume fraction $\phi = 0.63$) [44]. This appears inconsistent with experimental observations where the viscosity divergence occurs instead at the colloidal glass transition ($\phi = \phi_g = 0.58$) [45]. Accordingly it is necessary to develop a new theory, as outlined in Section 3, to describe flow curves at $\phi > \phi_g$. The hypothesis of the work on colloidal jamming reported here is that the proximity of this transition also affects flow properties in a window of densities *below* ϕ_g, to the extent that one should treat hydrodynamic forces as a perturbation to the dynamics of collective arrest, rather than vice versa.

6. CONCLUSION

Mode Coupling Theory (MCT) has had important recent successes, such as predicting, in advance of experiment, the re-entrant glass/gel nonergodicity curves that arise in colloidal systems with short range attractions [14–17].

Theoretical developments directly inspired by MCT now offer a promising framework for calculating the nonlinear flow behaviour of colloidal glasses and glassy liquids [21]. In fact, this offers the only current prospect for *quantitative* prediction of yield behaviour and nonlinear rheology in this or any other class of nonergodic soft materials. (Other work on the rheology of glasses [27,46] does not, as yet, offer quantitative prediction of experimental quantities.) While promising, many things are missing so far from the approach initiated in [21]: velocity fluctuations, hydrodynamic forces, anharmonicity in H etc., are all ignored. The fact that only shear thinning is predicted in this case is excusable.

The schematic work of Ref.[37] on shear thickening is preliminary, but interesting in that it suggests how new physics (beyond two-point correlations) may need to be added to MCT before the full range of observed colloidal flow behaviour is properly described. Hydrodynamic interactions, and perhaps velocity fluctuations, are certainly also important in some aspects of shear thickening, as discussed in Section 5, though we might hope that these do not dominate very close to the glass transition where the longest relaxation time is structural rather than hydrodynamic. Of course, even for systems at rest, it is known that some important physics is missing from MCT, in particular, the kinds of activated dynamics discussed in Section 2. These allow the system to move exponentially slowly despite being in a region of phase space where, according to MCT, it cannot move at all (see e.g. [8]). Qualitatively, stress-induced jamming seems a quite different phenomenon from this, although one cannot rule out a link of some sort (e.g. if stress switches off the activated processes [34]). Accordingly we can suspect that there are more things missing from MCT than just activated processes. In particular a more general treatment of anharmonic terms (or equivalently, a treatment of three-point and higher order correlations) may be required before one has a fully workable theory of sheared colloidal glasses.

REFERENCES

1. Cates M. E. and Evans M. R., "Soft and Fragile Matter: Nonequilibrium Dynamics, Metastability and Flow" IOP Publishing, Bristol (2000).
2. Pine D. J., "Light Scattering and Rheology of Complex Fluids Driven far from Equilibrium", in [1], pp.9–47.
3. Kob W., "Supercooled Liquids and Glasses", in [1], pp. 259–284.
4. Poon W. C. K., Starrs L., Meeker S. P., Moussaid A., Evans R. M. L., Pusey P. N. and Robins M. M., Faraday Discuss. 112 (1999) 143–154.
   Poon W. C. K., Renth F., Evans R. M. L., Fairhurst D. J., Cates M. E. and Pusey P. N., Phys .Rev. Lett. 83 (1999) 1239–1243.
5. Kroy K., et al., work in progress.
6. Götze W. and Sjoegren L., Rep. Prog. Phys. 55 (1992), 241–376.
7. Ramaswamy S., "Self-Diffusion of Colloids at Freezing", in "Theoretical Challenges in the Dynamics of Complex Fluids,", McLeish T. C. B., Ed., pp7-20, Kluwer, Dordrecht 1987.
8. Kawasaki K. and Kim B., J. Phys. Cond. Mat. 14 (2002) 2265–2273.
9. Cates M. E., cond-mat/0211066, to appear in Ann. Henri Poincare.
10. Garrahan J. P. and Chandler D., Phys. Rev. Lett. 89 (2002) 035704; J. Garrahan, this volume.
11. Weeks E. R., Crocker J. C., Levitt A. C., Schofield A., Weitz D. A., Science 287 (2000)

627.
12. Weitz D. A., this volume.
13. Götze W., "Aspects of Structural Glass Transitions", in "Liquids, Freezing and Glass Transition", Hansen J. P., Levesque D., and Zinn-Justin J. Eds (Les Houches Session LI), North Holland 1991, pp. 287–503. van Megen W., Underwood S. M. and Pusey P. N., Phys. Rev. Lett. 67 (1991) 1586–1589; van Megen W. and Underwood S. M., Phys. Rev. E. 49 (1994) 4206–4220; Götze W., J. Phys.: Cond. Matt. 11, A1-A45 (1999).
14. Bergenholtz J., and Fuchs, M., Phys. Rev. E 59 (1999) 5706–5715.
15. Dawson K., Foffi G., Fuchs M., Gotze W., Sciortino F., Sperl M., Tartaglia P., Voigtmann T., Zaccarelli E., Phys. Rev. E 63 (2001) 011401.
16. Fabbian L., Götze W., Sciortino F., Tartaglia P., and Thiery F., Phys. Rev. E 59 (1999) R1347–R1350. 1999; Sciortino F., this volume, Sciortino F., Nature Materials 1 (2003) 145–146.
17. Pham K. N., Puertas A. M., Bergenholtz J., Egelhaaf S. U., Moussaid A., Pusey P. N., Schofield A. B., Cates M. E., Fuchs M. and Poon W. C. K., Science 296 (2002) 104–106. Poon W. C. K., Pham K. N., Egelhaaf S. U. and Pusey P. N., J. Phys. Cond. Matt. 15 (2003) S269-S275.
18. Pham K. N., private communication.
19. Puertas A. M., Fuchs M. and Cates M. E., Phys. Rev. Lett. 88 (2002) 098301.
20. Sciortino F., Tartaglia P., Zaccarelli E., cond-mat/0304192.
21. Fuchs M. and Cates M. E., Phys. Rev. Lett. 89 (2002) 248303; Faraday Discussion 123 (2002) 267–286.
22. Indrani A. V. and Ramaswamy S., Phys. Rev. E 52 (1995) 6492–6496.
23. Cates M. E. and Milner S. T., Phys. Rev. Lett. 62 (1989) 1856–1859.
24. Miyazaki K., Reichman D. R., Phys. Rev. E 66 (2002) 050501(R).
25. Fuchs M. and Cates M. E., in preparation.
26. Bouchaud J.-P., J. Physique I 2 (1992) 1705–1713.
27. Fielding S. M., Sollich P. and Cates M. E., J. Rheol. 44 (2000) 323–369; Sollich P., Lequeux, F., Hebraud P. and Cates M. E., Phys. Rev. Lett. 78 (1987) 2020–2023. Sollich P., Phys. Rev. E 58 (1998) 738–759.
28. Barnes H. A. Hutton J. F. and Walters, K., "An introduction to rheology", Elsevier, Amsterdam 1989.
29. Götze W., Z. Phys. B 60 (1985) 195.
30. Fuchs M., Cates M. E., J. Phys. Cond. Mat. 15 (2003) S401.
31. Berthier L. and Barrat J. L., J. Chem. Phys. 116 (2002) 6228.
32. Laun H. M., J. Non-Newtonian Fluid Mec. 54 (1994) 87–108.
33. Bender J. and Wagner N. J., J. Rheol. 40 (1996) 889–916.
34. Head D. A., Ajdari A. and Cates M. E., Phys. Rev. E 64 (2001) 061509.
35. Bertrand E., Bibette J. and Schmitt V., Phys. Rev. E 66 (2002) 06040(R).
36. Cates M. E., Wittmer J. P., Bouchaud J.-P. and Claudin P., Phys. Rev. Lett. 81 (1998) 1841–1844. Liu A. J. and Nagel S. R., Nature 396 (1998) 21–22. Ball R. C. and Melrose J. R., Adv. Colloid Interface Sci. 59 (1995) 19–30.
37. Holmes C., Fuchs M. and Cates M. E., Europhys. Lett. 63 (2003) 240–246.
38. Frith W. J., d'Haene P., Buscall R. and Mewis J., J. Rheol40 (1996)531.

39. O'Brien V. T. and Mackay M. E., Langmuir 16(2000)7931.
40. Brady J. F., Curr. Opin. Colloid Interface Sci. 1 (1996) 472, and references therein.
41. Farr, Melrose, Ball Phys. Rev. E 55 (1997) 7203–7211.
42. Bergenholtz J., Brady J. F., Vicic M., J. Fluid Mech. 456 (2002) 239–275.
43. Ball R. C. and Melrose J. R., Adv. Colloid Interface Sci. 59(1995)19.
44. Brady J. F., J. Chem. Phys. 91 (1993), 3335–3341.
45. Segre P. N., Meeker S. P., Pusey P. N. and Poon W. C. K., Phys. Rev. Lett. 75 (1995) 958–961; Cheng Z., Zhu J., Chaikin P. M., Phan S.-E., and Russel W. B., Phys. Rev. E 65, 041405 (2002).
46. Berthier L., Barrat J.-L. and Kurchan J. Phys. Rev. E 61 (2000) 5464–5472.

Workshop participants list

- Berengere Abou, Université Paris VI Jussieu, Paris, France, *abou@ccr.jussieu.fr*;
- Jeferson Arenzon, IF-UFRGS Porto Alegre, Brazil, *arenzon@if.ufrgs.br*;
- Tomaso Aste, Australian National University Canberra, Australia, *tomaso.aste@anu.edu.au*;
- Daniel Bideau, Université de Rennes, France, *daniel.bideau@univ-rennes1.fr*;
- Jasna Brujic, University of Cambridge, Cambridge, UK, *jb333@phy.cam.ac.uk*;
- Gabriel Caballero, Université Paris VI Jussieu, Paris, France, *caballer@ccr.jussieu.fr*;
- Antonio Caiazzo, University of Naples, INFM, Italy, *antonio.caiazzo@na.infn.it*;
- Mike C. Cates, Edinburgh University, UK, *mec@ph.ed.ac.uk*;
- David Chandler, UCLA Berkeley, USA, *chandler@cchem.berkeley.edu*;
- Eric Clement, Université Paris VI Jussieu, Paris, France, *Eric.Clement@ccr.jussieu.fr*;
- Antonio Coniglio, University of Naples, Italy, *coniglio@na.infn.it*;
- Gianfranco D'Anna, EPFL Lausanne, Switzerland, *gianfranco.danna@epfl.ch*;
- Olivier Dauchot, CEA-SACLAY, Gif-sur-Yvette, France, *olivier.dauchot@cea.fr*;
- David Dean, Université de Toulouse, France, *dean@irsamc.ups-tlse.fr*;
- Lucilla de Arcangelis, Second University of Naples, INFM, Italy, *lucilla.dearcangelis@na.infn.it*;
- Antonio de Candia, University of Naples, INFM, Italy, *decandia@na.infn.it*;
- Emanuela Del Gado, Université de Montpellier, France, *emanuela.delgado@ldv.univ-montp2.fr*;
- Cristiano De Michele, University of Naples, INFM, Italy, *cristiano.demichele@na.infn.it*;
- Franco di Liberto, University of Naples, INFM, Italy, *franco.diliberto@na.infn.it*;
- Tiziana Di Matteo, Australian National University Canberra, Australia, *tiziana.dimatteo@anu.edu.au*;
- Annalisa Fierro, University of Naples, INFM, Italy, *fierro@na.infn.it*;
- Hans Fogedby, University of Aarhus, Denmark, *fogedby@phys.au.dk*;
- Silvio Franz, ICTP Trieste, Italy, *franz@ictp.trieste.it*;
- Juan P. Garrahan, Oxford University, UK, *j.garrahan1@physics.ox.ac.uk*;
- Chay Goldenberg, Tel-Aviv University, Israel, *chayg@eng.tau.ac.il*;
- David Head, University of Amsterdam, The Netherlands, *dhead@nat.vu.nl*;
- Yong Kheng Goh, Imperial College, London, UK, *yong.goh@ic.ac.uk*;
- Walter Kob, Université de Montpellier, France, *Walter.Kob@ldv.univ-montp2.fr*;

- Anke Lindner, Université Paris VI Jussieu, Paris, France, *lindner@ccr.jussieu.fr*;
- Hernan Makse, Levich Institute, New York, USA, *makse@levdec.engr.ccny.cuny.edu*;
- Francesco Mallamace, University of Messina, Italy, *mallamac@www.unime.it*;
- Anita Mehta, NCBS, Calcutta, India, *anita@bose.res.in*;
- Rene Mikkelsen, The Netherlands, *r.mikkelsen@tn.utwente.nl*;
- Alexander Morozov, University of Leiden, Leiden, The Netherlands, *morozov@lorentz.leidenuniv.nl*;
- Tom Mullin, University of Manchester, UK, *tom@reynolds.ph.man.ac.uk*;
- Sidney Nagel, University of Chicago, USA, *srnagel@uchicago.edu*;
- Mario Nicodemi, University of Naples, INFM and INFN, Italy, *Mario.Nicodemi@na.infn.it*;
- Giorgio Parisi, University of Rome, INFM and INFN, Italy, *Giorgio.Parisi@roma1.infn.it*;
- Massimo Pica Ciamarra, Texas University, Austin, USA, *massimo.picaciamarra@na.infn.it*;
- Felix Ritort, University of Barcelona, Spain, *ritort@ffn.ub.es*;
- Giancarlo Ruocco, University of Rome, INFM, Italy, *Giancarlo.Ruocco@roma1.infn.it*;
- Srikanth Sastry, University of Bangalore, India, *sastry@jncasr.ac.in*;
- Nicolas Sator, Université Paris VI Jussieu, Paris, France, *sator@lptl.jussieu.fr*;
- Matthias Schroeter, Texas University, Austin, USA, *schroeter@chaos.utexas.edu*;
- Francesco Sciortino, University of Rome, INFM, Italy, *francesco.sciortino@phys.uniroma1.it*;
- Mauro Sellitto, ICTP Trieste, INFM, Italy, *sellitto@ictp.trieste.it*;
- Jacco Snoeijer, University of Leiden, Leiden, The Netherlands, *snoeijer@lorentz.leidenuniv.nl*;
- H.Eugene Stanley, Boston University, USA, *hes@meta.bu.edu*;
- Attilio Stella, University of Padova, Italy, *stella@pd.infn.it*;
- Harry Swinney, Texas University, Austin, USA, *swinney@chaos.ph.utexas.edu*;
- Gilles Tarjus, Université Paris VI Jussieu, Paris, France, *tarjus@lptl.jussieu.fr*;
- Piero Tartaglia, University of Rome, INFM, Italy, *piero.tartaglia@phys.uniroma1.it*;
- Marco Tarzia, University of Naples, INFM, Italy, *marco.tarzia@na.infn.it*;
- Tamas Unger, Budapest University, Budapest, Hungary, *unger@maxwell.phy.bme.hu*;
- Paolo Verrocchio, Universidad Complutense de Madrid, Madrid, Spain, *paolo@lattice.fis.ucm.es*;
- David Weitz, Harvard University, Cambridge, USA, *weitz@deas.harvard.edu*;
- Marco Zannetti, University of Salerno, INFM, Italy, *Marco.Zannetti@na.infn.it*.

Index